Introduction to the Organic Nomenclature System

INTRODUCTION TO THE ORGANIC NOMENCLATURE SYSTEM

Rainer Roldan Fiscal

ARCLER
PRESS

www.arclerpress.com

Introduction to the Organic Nomenclature System

Rainer Roldan Fiscal

Arcler Press
224 Shoreacres Road
Burlington, ON L7L 2H2
Canada
www.arclerpress.com
Email: orders@arclereducation.com

ISBN: 978-1-77469-843-3 (Hardcover)

ABOUT THE AUTHOR

Rainer Roldan Fiscal is an Associate Professor V at the Laguna State Polytechnic University Siniloan Campus, Philippines. He graduated Bachelor of Secondary Education major in General Science, Master of Arts in Teaching Science and Technology, and Doctor of Philosophy in Education major in Educational Leadership and Management. He teaches Biological Science, Physical Science, and Research courses in the undergraduate and graduate teacher education programs. He presented research papers at national and international conferences in the Philippines and other countries like Malaysia, Thailand, and Vietnam. He published research articles in international journals and has citations.

TABLE OF CONTENTS

LIST OF FIGURES

LIST OF ABBREVIATIONS

ATP	Adenosine Triphosphate
BHB	Beta-Hydroxybutyrate
C_2H_4	Ethylene
C_3H_6	Propylene
C_4H_{10}	Butane
C_6H_6	Benzene
CH_4	Methane
CIP	Cahn-Ingold-Prelog
DHA	Dihydroxyacetone
DNA	Deoxyribonucleic Acid
FDA	Food and Drug Administration
FGs	Functional Groups
HCN	Hydrogen Cyanide
IUPAC	International Union of Pure and Applied Chemistry
$KMnO_4$	Potassium Permanganate
LDA	Lithium Diisopropylamide
$NaNH_2$	Sodium Amide
NH_3	Ammonia
OsO_4	Osmium Dioxide
PAH	Polycyclic Aromatic Hydrocarbons
pH	Potential of Hydrogen
RCOOH	Carboxylic Acids
RNA	Ribonucleic Acid
SAR	Structure-Activity Relationship
SPF	Sun Protection Factor
UV	Ultraviolet

PREFACE

The IUPAC nomenclature system involves a list of scientific rules that organic chemists developed and use to circumvent issues brought on by arbitrary nomenclature. With these guidelines and a structural formula, you can give each compound a unique name. To make communication easier, the IUPAC system of nomenclature aims to establish a global standard for naming compounds. The system's objective is to associate each name with a distinct and unambiguous structure and provide each structure with a distinct and clear name.

The IUPAC nomenclature relies on naming the molecule's longest single-bond carbon chain, whether it is a ring or a continuous chain. Prefixes and suffixes are used to indicate any deviation—multiple bonds and atoms apart from carbon and hydrogen—according to a certain order of priorities. This volume mentions that the sheer number of new discoveries of organic elements necessitated such a systematic approach because the trivial nomenclature of organic compounds was extremely inconvenient. Nevertheless, chemists do not always adhere to the IUPAC nomenclature guidelines because some compounds have extremely lengthy and tiresome names. The names given to these compounds are less important.

The ideal situation is for every possible organic compound to have a name whereby a unique structural formula can be constructed. IUPAC names are also used in inorganic chemistry. The official IUPAC system for identifying compounds is not always followed in practice, with the exemption of situations whereby a compound must be provided an absolute definition.

Alkanes, which are molecules comprising carbon and hydrogen that are only related by single bonds, are a family of saturated hydrocarbons. These molecules can be linear or alicyclic or in continuous chains known as cyclic or alicyclic. The basic names of organic compounds are alkanes and cycloalkanes. Starting with the five-carbon alkane, the number of Greek and Latin prefixes demonstrate the number of carbons within the chain "cyclo" is used to identify rings. Each apex of a ring is represented by carbon and the number of hydrogens needed to complete its valence in the geometric symbols.

The functional group with the highest priority is indicated by a suffix when naming molecules that contain one or more of the Group A functional groups; the others are shown by prefixes, with a need comparable to some other substituents. Alkene and alkyne, two functional groups of Group B, are regarded as having equal priority: The numbering direction in a molecule with double and triple bonds is determined by the one that is closest to the chain's end. The double bond takes the lower number if each would contain the same position number. Due to alphabetization, "ene" appears before "yne" in the name.

Compounds that are obtained from benzene and comparable ring systems are referred to as "aromatic." The procedure is the same as that described above for aliphatic nomenclature: determining the parent ring's root name; defining the name, position, and priority number of substitutes; and putting the name together alphabetically. Aliphatic and aromatic nomenclature share the same functional group priorities. Ketones are a distinct class of aromatic compounds in which the carbonyl is linked to one or more benzene rings. The scope and nature of the group contrary to carbonyl determine the prefix with which these compounds are referred to as "phenones."

CHAPTER

1 The IUPAC Nomenclature System

Contents

The IUPAC nomenclature system (International Union of Pure and Applied Chemistry) is a set of rules and conventions for naming chemical compounds. It provides a standardized approach to ensuring clear and unambiguous communication among scientists. The IUPAC nomenclature system is critical for chemists because it allows them to accurately describe compounds and facilitates knowledge sharing across fields and regions.

The IUPAC nomenclature system's primary goal is to assign unique and systematic names to chemical compounds based on their molecular structure and composition. Chemists can generate names that reflect the characteristics of the compound and help identify its properties and behavior by following a set of well-defined rules. Organic and inorganic molecules, ions, coordination compounds, polymers, and biochemicals are all covered by this system (Ehrenhauser, 2015).

Various principles, such as prioritizing functional groups, indicating the presence of substituents, and using prefixes and suffixes to denote the number and type of atoms in a compound, are incorporated into the IUPAC nomenclature system. It also provides guidelines for naming complex compounds and is constantly evolving to accommodate new chemistry discoveries and advancements.

The IUPAC nomenclature system promotes consistency and clarity in scientific literature, research papers, patents, and chemical databases. It allows chemists to communicate effectively, facilitates information exchange, and contributes to the overall advancement of chemistry as a scientific discipline. Important aspects of the IUPAC nomenclature system include the following areas (Johnson, 2010).

1.1. NOMENCLATURE FOR ORGANIC CHEMISTRY

To describe organic chemical compounds, the International Union of Pure and Applied Chemistry (IUPAC) recommends using organic chemistry nomenclature in chemical nomenclature. It is published by the Organic Chemistry Nomenclature.

Any conceivable organic compound should ideally have a name that can be translated into a precise structural formula. Inorganic chemistry also employs the IUPAC nomenclature.

In practice, the official IUPAC recommendations for naming are not always followed, except when a compound requires an unambiguous and absolute definition to avoid lengthy and tedious names in normal communication. IUPAC names are sometimes shorter than older names. Ethanol, for example, is a shorter name than ethyl alcohol.

They are easier to understand for relatively simple molecules than non-systematic names that must be memorized or read over. The common or trivial name, on the other hand, is frequently much shorter and clearer, and thus preferred. These bizarre names are frequently derived from a single compound. Long names may also be more difficult to remember than structural formulas (Ehrenhauser, 2015).

1.1.1. The Organic Chemistry Nomenclature

Every single thing in the universe has a name that identifies it uniquely. When organic chemistry was new and there were few known compounds, compounds were named after their sources or discoverers. These are the so-called trivial or common names.

The IUPAC developed a set of guidelines for the systematic nomenclature of organic compounds, which are periodically revised. The three main components of the IUPAC nomenclature for organic compounds are the stem name, prefix, and suffix.

It is worth noting that the presence of preferred IUPAC names does not preclude the use of alternative names to account for a specific context or to highlight structural traits that a group of compounds shares. The term "preferred IUPAC nomenclature" refers to a set of preferred IUPAC names. Any name, even if it is not a preferred IUPAC name, is acceptable as a general IUPAC name in the context of a "general IUPAC nomenclature" as long as it is clear and follows the guidelines of the IUPAC recommendations here (Johnson, 2010).

1.1.2. Trivial Nomenclature System

Organic compounds are named using a non-systematic method known as trivial nomenclature. There are no specific guidelines for writing compound petty naming. This system simplifies the names of organic compounds. Toluene, acetic acid, and phenol are just a few examples.

When compared to the corresponding IUPAC nomenclature of the same compounds, trivial nomenclature names are frequently much shorter and simpler. As a result, this system is still relevant today.

The following example demonstrates the relative ease with which compounds can be named: Tartaric acid is a type of carboxylic acid found in tamarind, according to the trivial system. Tartaric acid's IUPAC nomenclature is 2,3-dihydroxy-1,4-butanedioic acid (Ehrenhauser, 2015).

1.1.3. Disadvantages of the Trivial Nomenclature System

The disadvantages of the simple system for naming organic compounds are listed below.

- A single compound may have several meaningless names.
- The Trivial nomenclature system only applies to a small number of compounds within each distinct group, as evidenced by the fact that Phenol also goes by hydroxybenzene and carbolic acid. Formic acid and acetic acid, the first two members of the carboxylic acid group, are good examples of this. However, there are no simple names for carboxylic acids with more atoms.
- There is no specific set of rules for complex compound nomenclature in the trivial system (Ehrenhauser, 2015).

1.2. NOMENCLATURE OF ORGANIC CHEMICALS

Because of the increasing number of organic compounds discovered every day, as well as the fact that many of these compounds are isomers of other compounds, a systematic nomenclature system is required. Each distinct compound requires its own name, just as each distinct compound has its own molecular structure that can be identified by a structural formula.

As organic chemistry advanced and expanded, many compounds were given unimportant names; these names are now well-known and understood. Methane, butane, and other similar substances (Figure 1.1).

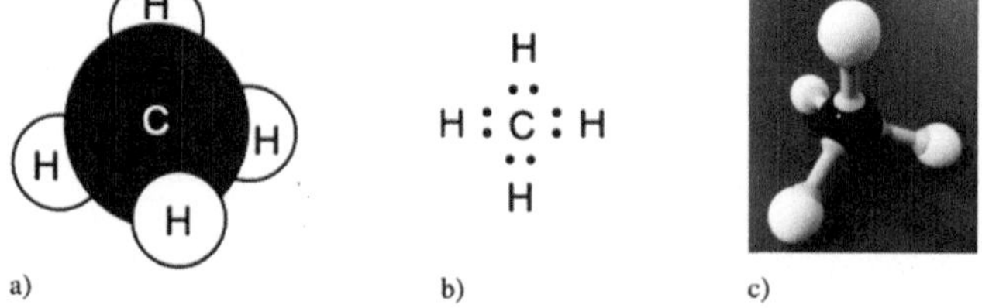

Figure 1.1. A methane compound.

Source: *Patron Emelie, (2021); Scientific Figure on ResearchGate.*

These common names frequently have roots in scientific history and the natural origins of specific compounds, but their relationships to one another are arbitrary and lack any systematic or logical principles.

1.2.1. The IUPAC Approach to Systematic Nomenclature

A rational nomenclature system should achieve at least two objectives. It should begin by describing the distinctive lattice of chains and rings that hold the carbon atoms of a particular compound together. Second, it must identify any functional groups found in the compound. Because hydrogen is a common component of organic compounds, its amount and location can usually be estimated using carbon's tetravalency (Ehrenhauser, 2015).

To avoid problems caused by arbitrary nomenclature, organic chemists created and use the IUPAC nomenclature system. Given these guidelines and a structural formula, it should be possible to generate a unique name for each distinct compound. Similarly, given an IUPAC name, one should be able to generate a structural formula. A typical IUPAC name will have three key characteristics:

- The names of the hydrogen-free substituent groups that complete the molecular structure.
- A suffix or other element(s) designating functional groups that may be present in the compound.
- A root or base designating a major chain or ring of carbon atoms found in the molecular structure.

As an introduction to the IUPAC nomenclature system, we will begin by looking at compounds with no discernible functional groups. Only sigma-bonded carbon and hydrogen atoms are present in these substances (Ehrenhauser, 2015).

1.2.2. Alkanes

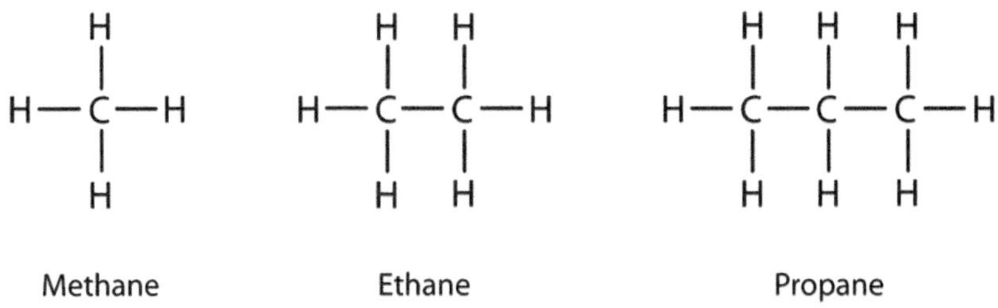

Figure 1.2. A representation of symmetry alkanes.

Source: *Gunes Aysun, Togan Muge, Celik Feriha, & Cangul Ismail Naci, (2019); Scientific Figure on ResearchGate.*

Hydrocarbons without double or triple-bond functional groups are classified as alkanes or cycloalkanes depending on whether the carbon atoms in the molecule are arranged only in chains or also in rings. Despite the lack of functional groups, these hydrocarbons serve as the structural foundation for the placement of functional groups in other classes of compounds, making them an excellent place to start when researching and naming organic compounds. Alkanes and cycloalkanes are both members of the aliphatic group of substances. To put it simply, aliphatic substances are those that lack aromatic rings in their molecular structure (Figure 1.2).

The table below lists the IUPAC names for simple continuous-chain alkanes between C-1 and C-10. The typical "ane" suffix classifies these compounds as alkanes. Longer chain alkanes are well known, and their names appear frequently in textbooks and references. The names methane through decane should be memorized because they serve as the foundation for many IUPAC names. Fortunately, names for chains with five or more carbon atoms have consistent numerical prefixes (Johnson, 2010).

Name	**Molecular Formula**	**Structural Formula**	**Isomers**
Methane	CH_4	CH_4	1
Ethane	C_2H_6	CH_3CH_3	1
Propane	C_3H_8	$CH_3CH_2CH_3$	1
Butane	C_4H_{10}	$CH_3CH_2CH_2CH_3$	2
Pentane	C_5H_{12}	$CH_3\ (CH_2)_3CH_3$	3
Hexane	C_6H_{14}	$CH_3\ (CH_2)_4CH_3$	5
Heptane	C_7H_{16}	$CH_3\ (CH_2)_5CH_3$	9
Octane	C_8H_{18}	$CH_3\ (CH_2)_6CH_3$	18
Nonane	C_9H_{20}	$CH_3\ (CH_2)_7CH_3$	35
Decane	$C10H_{22}$	$(CH_2)_8CH_3$	75

Key points to note in groups of compounds.

i. All of these equations follow the CnH_2n behavioral trends including

ii. The structures and formulas of these alkanes uniformly increase by a CH_2 increment.

iii. Homologous refers to this type of variation occurring consistently 2 rules. Furthermore, this is the highest H/C ratio that a stable hydrocarbon can have.

iv. These compounds are called saturated (with hydrogen) because their H/C ratio is at its highest.

There are alkane isomers present, beginning with butane (C_4H_{10}) and increasing in number as alkanes become larger. C_6H_{14} isomers, for example, have five isomers, as shown below.

a. n-hexane;

b. 2-methylpentane;

c. 3-methylpentane;

d. 2,2-dimethylbutane; and

e. 2,3-dimethylbutane.

Despite having the same molecular formula, only the first of these two compounds is known as hexane. So, how do we find the others?

As previously stated, the IUPAC system requires names first for simple unbranched chains, followed by names for any possible simple alkyl groups. The table below contains some examples of common alkyl groups. When naming groups, the "yl" suffix replaces the "ane" suffix. The symbol R represents a general (unspecified) alkyl group.

The IUPAC Alkane Nomenclature Rules are as follows attached to the chains:

1. Find and identify the longest continuous carbon chain.
2. Identify the groups that comprise this chain.
3. Begin sequentially numbering the chain at the location of each substituent group.
4. Put the name together, listing the groups alphabetically using the full name (cyclopropyl, for example, comes before isobutyl). end closest to a substituent group.
5. Use the appropriate number and name to indicate the prefixes di, tri, tetra, and so on, which are used to designate multiple groups of the same kind, are ignored when alphabetizing.

The names fluoro (F-), chloro (Cl-), bromo (Br-), and iodo (I-) make it easy to accommodate halogen substituents. For example, 1-bromo-3-methylbutane would be denoted as $(CH_3)_2CHCH_2CH_2Br$. If the halogen is bonded to a simple alkyl group, the name "alkyl halide" may be used instead. As a result, C_2H_5Cl is also known as ethyl chloride or chloroethane (a two-carbon chain does not require a locator number). The full names of halogenated alkyl substituents like bromomethyl ($BrCH_2$) and trichloromethyl (CCl_3-)

can be more carbon atom rings. The simplest examples of this class consist of one unsubstituted listed and sorted alphabetically (Ehrenhauser, 2015).

1.2.3. Cycloalkanes

Cycloalkanes are compounds made up of one carbon ring and form a homologous series with unbranched alkanes. The IUPAC names of the first five members of this series are listed in the table below. The final (yellow-shaded) column contains the general formula for any size cycloalkane. To become a cycloalkane, a simple unbranched alkane must lose two hydrogen atoms, one from each end of the chain. As a result, CnH_2n is the general formula for a cycloalkane containing n carbons. Despite having two fewer hydrogens than the equivalent alkane, cycloalkanes are considered to have a saturated hydrogen content because each carbon is bonded to four other atoms (Johnson, 2010).

The naming conventions for substituted cycloalkanes are very similar to those for branched alkanes. The numbering system is where the rules and procedures diverge the most. Because rings, unlike chains, have no ends, all of the carbons in a ring are equal, so the numbering starts with a substituted ring atom.

1.2.3.1. IUPAC Rules for Cycloalkane Nomenclature

1. When naming a monosubstituted cycloalkane, the substituent group is named first, followed by the ring (see table above). A location number is not required.
2. If the alkyl substituent is substantial and/or complex, the ring may be referred to as a substituent group on an alkane.
3. If the ring contains two different substituents, they are listed alphabetically, with carbon #1 assigned to the first cited substituent. The next step is to count the ring carbons in either a clockwise or counterclockwise direction, giving the second substituent the lowest possible location number.
4. If there are multiple substituents in the ring, they are listed alphabetically. The substituents are assigned location numbers, counting clockwise or counterclockwise so that one is at carbon #1 and the others have the lowest numbers.
5. Create the name by listing groups alphabetically and assigning a location number to each group (if there are two or more). The

prefixes di, tri, tetra, and so on, which are used to designate multiple groups of the same kind, are ignored when alphabetizing (Johnson, 2010).

Examples of how to use these rules to name substituted cycloalkanes.

Small rings with three and four members, in particular, suffer from significant angle strain as the sp^3 carbon bond angles shift from the ideal 109.5° to 60° and 90°, respectively. This angle strain frequently increases the chemical reactivity of these compounds, resulting in ring cleavage products. It is also critical to understand that, with the exception of cyclopropane, cycloalkyl rings are not planar (flat).

Hydrocarbons with two or three rings are commonly referred to as bicyclic compounds, tricyclic compounds, and polycyclic compounds. The H/C ratios of such compounds decrease as the number of rings in their molecular formulas increases. For instance, $CnH_{(2n + 2 - 2m)}$ is the general formula for a hydrocarbon composed of n carbon atoms linked to m rings. The structural relationships between the rings in a polycyclic compound can change. They could be independent and distinct, or they could share one or two atoms (Ehrenhauser, 2015).

1.2.4. Alkenes and Alkynes

Alkenes and alkynes are hydrocarbons with functional groups composed of carbon-carbon double bonds and carbon-carbon triple bonds, respectively. The molecular formulae of these unsaturated hydrocarbons contain multiple functional group bonds:

$R-CH_2-CH_2-R$ CnH_{2n+2} alkane This H/C ratio is the highest for a given number of carbon atoms.

R-CH=CH-R CnH_{2n} alkene Each double bond has two fewer hydrogen atoms.

R-C-C-R CnH_{2n-2} alkyne Each triple bond has four fewer hydrogen atoms.

As previously stated, the molecular formula of a hydrocarbon indicates the structural possibilities it may represent. As an example, consider compounds with the formula C_5H_8. Pentane, a five-carbon alkane, has the chemical formula C_5H_{12}, so the difference in hydrogen content is four. Because of this distinction, such compounds can have two rings, a triple bond, two double bonds, a ring, or three rings.

1.2.4.1. The IUPAC Nomenclature Alkene and Cycloalkene Rules

1. The ene suffix (ending) denotes an alkene or cycloalkene.
2. The two carbon atoms of the double bond must be present in the longest chain of the root name.
3. The root chain must be numbered beginning with the end closest to a carbon atom in a double bond. If the double bond is in the middle of the chain, the nearest substituent rule is used to determine where the numbering begins.
4. The double bond locator is the smaller of the two numbers that designate the double bond's carbon atoms. To indicate the number of double bonds present, the compound is given a name with a diene, triene, or equivalent prefix, and each double bond is assigned a locator number.
5. The double-bond carbons in cycloalkenes are assigned to ring positions #1 and #2. To determine which of the two is number one, apply the nearest substituent rule.

The vinyl group $H_2C=CH$ group $H_2C=CH-CH_2$- Allyl is an example of a double-bonded substituent group.

1.2.4.2. IUPAC Alkynes Nomenclature Rules

1. The yne suffix (ending) denotes an alkyne or cycloalkyne.
2. The two carbon atoms in the triple bond must be present in the longest chain of the root name.
3. The root chain must be numbered beginning with the end closest to a carbon atom in a triple bond. If the triple bond is in the middle of the chain, the nearest substituent rule is used to determine where the numbering begins.
4. The triple bond locator is the smaller of the two numbers that designate the triple bond's carbon atoms.
5. A locator number must be assigned to each multiple bond that exists. Although double bonds come before triple bonds in the IUPAC name, the chain is numbered from the end that is closest to multiple bonds, regardless of the type of bond (Ehrenhauser, 2015).
6. Because the triple bond is linear, it can only fit in rings with more than ten carbons. In simple cycloalkynes, the triple bond carbons

are assigned ring positions #1 and #2. To determine which of the two is number one, apply the nearest substituent rule.

7. Triple-bonded substituent groups include the following:

- consider the HC-C- ethynyl group.
- The HC-C-CH_2-propargyl group

1.2.5. Benzene-Derived Compounds

Substituted benzene ring compounds have a less systematic nomenclature than alkanes, alkenes, and alkynes. The combined names shown below show how some mono-substituted compounds are named by prefixing "benzene" with a group name. The majority of these compounds, however, are identified by unique names. Memorization is the only viable option for mastering these names.

Phenyl and benzyl, two commonly used substituent groups containing benzene rings, are denoted by the letters Ph- and Bn-, respectively (Johnson, 2010).

When there are multiple substituents in a benzene ring, their relative positions must be indicated either by numbering the ring carbons or by some other notation. In the case of disubstituted benzenes, the prefixes ortho, meta, and para are frequently used to denote a 1,2, 1,3, or 1,4 relationships, respectively. Some disubstituted toluenes, such as xylene, cresol, and toluidine, use their singular names to identify their isomers, which are typically prefixed with ortho, meta, or para. Some disubstituted benzenes (such as salicylic acid and resorcinol) have distinct names for different isomers. Finally, if there are three or more substituent groups, the ring is numbered in such a way that the substituents have the lowest possible numbers. The substitutes are listed alphabetically in the final name. When the substitution is symmetrical, the numbering is alphabetical.

1.3. SUBSTITUTIVE NOMENCLATURE

Substituted nomenclature is a key component of the IUPAC Nomenclature System, which is used to name organic compounds. It provides a systematic method for recognizing and naming any substituents or functional groups found in a molecule. Understanding the rules of substitutive nomenclature allows chemists to accurately communicate the structure and composition of organic compounds. In this discussion, we will look at the main ideas and guidelines of substitutive nomenclature.

The idea behind substitutive nomenclature is that the substituent or functional group of a molecule is the primary component, while the rest of the molecule is considered the parent or base structure. Following the name of the substituent is the name of the parent structure. Chemists can use this method to quickly identify and classify compounds based on their functional groups.

The first step in substitutive nomenclature is to recognize and name the organic molecule substituents. A substituent is a group of atoms that takes the place of one or more hydrogen atoms in a hydrocarbon chain. It is frequently indicated by a prefix attached to the parent structure. The prefix indicates the number of carbon atoms in the substituent and can also include numerical prefixes such as "di-" (two), "tri-" (three), and so on to denote multiple substituents of the same type.

The names of the substituents are derived from hydrocarbons with the same number of carbon atoms. A "methyl group" is formed by one carbon atom; two carbon atoms form an "ethyl group," three carbon atoms form a "propyl group," and so on. Names of substituents derived from hydrocarbons with four or more carbon atoms show similar patterns (Ehrenhauser, 2015).

When a molecule has multiple substituents, they are named alphabetically. When there are multiple identical substituents, numerical prefixes are used to denote multiplicity. Trimethyl, a molecule with three methyl groups, and dimethyl, a molecule with two methyl groups, are two examples.

The name of the parent structure is determined by the length of the longest continuous carbon chain or backbone in substitutive nomenclature. The chain in question is the parent chain. The functional groups or substituents are named in relation to the parent chain, which also serves as the foundation for the compound's name.

The positions of the substituents on the parent chain are denoted by numbers. The numbering begins at the end closest to the first substitute. The goal is to give the substituents the lowest possible number. If multiple substituents of the same type exist, their positions are separated by hyphens. Commas separate the position numbers from one another. In a molecule with two methyl groups attached to the parent chain, for example, the positions would be 1,1-dimethyl (Johnson, 2010).

When there are several different substituents, the substituents are numbered in such a way that they receive the smallest possible combination of numbers. Commas separate the numbers, which are listed in ascending order. For example, the name 2-methyl-4-ethyl refers to a molecule that has a methyl group at position 2 and an ethyl group at position 4.

When a functional group is present in the parent structure, the suffix designating the functional group replaces the "-e" ending of the corresponding hydrocarbon in the name. Some common functional group suffixes are "-ol" for alcohols, "-one" for ketones, "-al" for aldehydes, "-oic acid" for carboxylic acids, and "-amine" for amines.

In addition to these general rules, there are additional rules for dealing with more complicated compounds such as cyclic structures, fused rings, and bridged compounds. In these circumstances, careful consideration of the parent structure, substituents, and their positions is required to ensure a precise and unambiguous name.

For organic compounds, the IUPAC Nomenclature System heavily relies on substitutive nomenclature. It provides a methodical approach to naming compounds by prioritizing functional groups or substituents. Chemists can effectively communicate the structure and makeup of organic molecules by adhering to the rules of substitutive nomenclature in order to accurately understand and communicate within the field of chemistry.

1.4. FUNCTIONAL GROUPS

A functional group is defined as a "specific grouping of elements for which the distinct chemical reactions of these molecules are responsible."

Chemical reactions involving two molecules with different sizes but the same functional groups will be similar or identical. When a molecule has a functional group, it means that its behavior and chemical reactions can be systematically predicted.

Chemical synthesis—the deliberate execution of chemical reactions to produce a specific compound—can be designed by understanding the properties of various functional groups (Ehrenhauser, 2015).

1.4.1. The Function of Functional Groups in Biological Chemistry Working Groups

The way functional groups participate in a chemical reaction can be altered further with the help of other functional groups. These groups can also be combined to form new ones.

Some examples of carbon-based functional groups are provided below. As a result, functional groups are moieties that have their own distinct characteristics and properties apart from the molecule to which they are attached.

- Covalent bonds connect the atoms of these groups and the group as a whole to the molecule.
- Ionic charge can be found on some functional groups, as seen in carboxylate salts containing the -COO- ionic group. In the case of polymers, the functional groups are typically attached to the nonpolar core of the carbon atoms in each repeating unit of the corresponding polymer, imparting specific chemical properties to the carbon chain.
- When these groups combine with molecules, the molecules are transformed into polyatomic ions or complexions.
- In a coordination complex, a ligand is a functional group that is bound to the central atom.

The presence of functional groups in a molecule influence both its solubility and its proclivity to form complexes. If the functional groups of the solute and the solvent interact well, solubility increases. Sugar, for example, dissolves easily in water because it contains both the -OH (hydroxyl) group and water.

Polarity is created when a less electronegative atom or molecule is attached to a highly electronegative functional group, allowing the initially nonpolar molecule to dissolve in water or other aqueous environments.

1.4.1.1. Nomenclature of Common Functional Groups

This section contains a list of the most common functional groups, as well as the prefix and suffixes that must be used in their nomenclature. A brief summary of the organizational structures of each of these groups is also provided.

Hydrocarbons:

- The letter R stands for alkanes, alkenes, and alkynes (as well as benzene derivatives on occasion). These groups are also known as hydrocarbyl groups because they only contain carbon and hydrogen atoms. However, the number of double or triple bonds formed between the two carbon atoms may differ.
- The type of carbon-carbon bond influences the reactivity of these groups. Some groups have unique names because they contain a

long, branched, or ring-structured alkane. Bornyl and cyclohexyl are two examples.

- The hydrocarbon functional groups could have an ionic charge. Carbocations refer to positively charged structures, whereas carbanions refer to negatively charged hydrocarbons (Johnson, 2010).

Haloalkanes:

- Haloalkanes, also known as alkyl halides, are carbon-halogen functional groups. The prefix 'Halo-' is used to identify a halogen.
- The prefix fluoro, for example, can be used to refer to the chemical CH_3F as fluoromethane. The suffix "halide" is used to identify a halogen. Using the suffix fluoride, for example, the same substance, fluoromethane (CH_3F), can also be referred to as methyl fluoride.
- The halogen determines the strength and stability of the carbon-halogen bond. Alkyl iodides, for example, have a weak carbon-iodine bond, whereas alkyl fluorides have a strong and stable carbon-fluorine bond.
- With the exception of these alkyl fluorides, all alkyl halides are easily subject to nucleophilic substitution or elimination reactions.

Oxygen-Containing Functional Groups:

The electron donating effect of sp^3 hybridized oxygen in alcohols, as opposed to the electron withdrawing effect of sp^2 hybridized oxygen in carbonyl groups containing a carbon-oxygen double bond, explains why the properties of functional groups containing a carbon-oxygen bond are entirely dependent on the hybridization of the carbon-oxygen bond (Ehrenhauser, 2015) (Figure 1.3).

It should be noted that the above list only includes a few common functional groups that contain the carbon-oxygen bond. Many unusual groups with complicated chemical compositions, such as acetal groups (RCH(OR')(OR'') or ketal groups (RCH(OR')(OR'')R. Nitrogen-containing functional groups. Carbon-oxygen bonds may also be found in nitrogen-containing substituent groups.

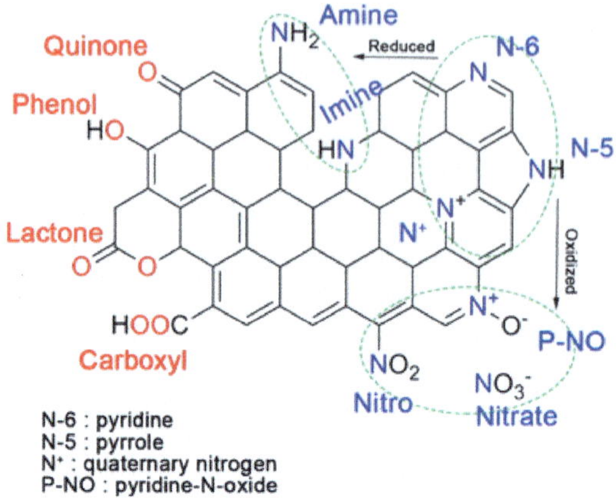

Figure 1.3. Types of oxygen-containing and nitrogen-containing functional groups.

Source: *Li Yuran, Lin Yuting, Cheng Chungui, Hao Junke, & Zhu Tingyu, (2020); Scientific Figure on ResearchGate.*

For example, the amide functional group has the formula R-(CO)-NR_2, indicating that it contains a carbonyl carbon bonded to a nitrogen atom, which is then bonded to two additional alkyl groups. The table below lists some of the most common nitrogen-containing functional groups, as well as the suffixes that make up their nomenclature. It should be noted that the pyridine derivatives with the formula RC_5H_4N, contain a large number of nitrogen-containing functional groups (Johnson, 2010).

1.5. HETEROCYCLIC COMPOUND NOMENCLATURE

Organic compounds with heterocyclic rings contain carbon atoms as well as atoms from other elements such as sulfur (S), nitrogen (N), oxygen (O), and other non-carbon atoms. The IUPAC Nomenclature System provides guidelines for naming heterocyclic compounds based on their structure and chemical makeup. In this discussion, we will look at the main tenets and rules for heterocyclic nomenclature.

One of the fundamental components of heterocyclic nomenclature is the identification of the heteroatom, the non-carbon atom presents in the ring structure. The heteroatom is frequently identified by an element symbol (e.g., N, O, or S) and is frequently numbered using the lowest locant. To help identify its position and ensure a distinct naming scheme, each atom in the ring is assigned a locant, which is a number.The number of carbon atoms in the heterocyclic ring is also an important factor in the naming process. Monocyclic heterocyclic rings have a single ring, whereas fused heterocyclic rings have two or more rings with shared carbon atoms. To name monocyclic heterocycles, the prefix "a" (for azacycles), "o" (for oxacycles), or "th" (for thioacycles) that corresponds to the number of carbon atoms in the ring is added before the parent hydrocarbon name. For example, pyrrole is a five-membered heterocyclic ring with nitrogen, whereas pyran is a six-membered ring with oxygen (Ehrenhauser, 2015).

The names of fused heterocycles are formed by combining the names of the individual rings with the appropriate prefixes. While the prefix "dehydro" indicates that the parent hydrocarbon contains unsaturated double bonds, the prefix "perhydro" indicates that all of the parent hydrocarbon's double bonds are saturated. Beginning at one of the fusion points, the numbering of fused rings extends through both rings.

Substituents can also alter the ring structure of heterocyclic compounds. The same principles that govern organic compound nomenclature apply to naming substitutes. Locants for the substituents are assigned based on their positions in relation to the heteroatom or any other functional groups present. If there are multiple substitutes, their locants are separated by commas and alphabetically listed.

Furthermore, heterocyclic compounds can exhibit isomerism, which occurs when two or more compounds with the same molecular formula differ in spatial orientations or structural arrangements. Isomerism in heterocyclic compounds can be caused by a variety of ring sizes, heteroatom positions, or the presence of multiple rings. The IUPAC nomenclature system includes specific guidelines for distinguishing between isomers (Johnson, 2010).

In some cases, the ring structure of a heterocyclic compound can accommodate multiple heteroatoms. To denote the locations of the heteroatoms, the names of the individual heterocycles in these compounds are prefixed with the appropriate locants. If multiple heteroatoms of the same element exist, their positions are indicated by the prefixes "di," "tri," and so on. The heteroatoms are organized alphabetically.

Stereochemistry is also involved in heterocyclic nomenclature. If stereoisomerism, such as cis-trans isomerism or optical isomerism, is present, any necessary descriptors are added to the name of a heterocyclic compound. Prefixes such as "cis," "trans," "R," or "S," which follow stereochemical rules, are used to indicate the stereochemical configuration of the substituents or heteroatoms within the ring. Finally, heterocyclic nomenclature refers to the systematic naming of organic compounds with ring structures containing atoms other than carbon. The heteroatom must be identified, the ring atoms numbered, the rings classified as monocyclic or fused, and the substituents assigned locants. The nomenclature of heterocyclic compounds is heavily influenced by stereochemistry and isomerism. Chemists can communicate clearly and consistently when discussing and naming heterocyclic compounds by following the rules of the IUPAC Nomenclature System.

1.5.1. Heterocyclic Compounds

The heterocyclic compounds are most likely the largest and most diverse family of organic compounds. After all, any carbocyclic compound, regardless of structure or functionality, can theoretically be transformed into a collection of heterocyclic analogs by substituting one or more of the ring carbon atoms with a different element. Even if we limit our consideration to oxygen, nitrogen, and sulfur (the most common heterocyclic elements), the permutations and combinations of such a replacement are numerous (Ehrenhauser, 2015).

1.5.2. Nomenclature

Nomenclature Developing a systematic nomenclature system for heterocyclic compounds has been a difficult task that has not always been successful. Many heterocycles, particularly amines, were discovered early on and given trite names that are still used today. The chart below depicts a few of these monocyclic compounds, with the systematic name based on the Hantzsch-Widman system provided beneath the common (trivial) name in blue. The operating instructions for this system will be provided later. After learning these common names, the majority of students will have a solid foundation in nomenclature (Figure 1.4).

The use of an elemental prefix for the heteroatom, followed by the appropriate carbocyclic name, results in a simple but limited nomenclature system. The table below contains a brief list of a few popular prefixes, with priority increasing from right to left. Examples of this nomenclature include ethylene oxide as 1-oxa-4-azacyclopropane, furan as oxacyclopenta-2,4-diene, pyridine as azabenzene, and morpholine as 1-oxa-4-azacyclohexane (Johnson, 2010).

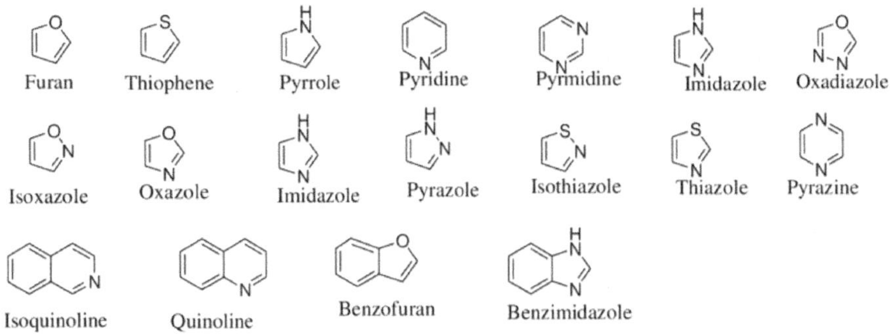

Figure 1.4. Structures of some heterocyclic compounds.

Source: *Kaur Ramandeep, Taheam Nisha, Sharma Anil Kumar, & Kharb Rajeev, (2013); Scientific Figure on ResearchGate.*

The Hantzsch-Widman system provides a more organized method of naming heterocyclic compounds that do not rely on previously used carbocyclic names. In this structure, the same hetero atom prefix from above is used (without the final "a"), followed by a suffix designating ring size and saturation. As shown in the table below, each suffix consists of a ring size root (blue) and an ending that indicates the degree of ring unsaturation. It is critical to understand that the unsaturated suffix applies to rings with the greatest number of non-cumulated double bonds, whereas the saturated suffix applies only to completely saturated ring systems. A prefix such as "dihydro" or "tetrahydro" is required for systems with a lower level of unsaturation.

- Despite the overall systematic design of the Hantzsch-Widman system, a number of modifications and exceptions have been added to account for inconsistencies with previous usage. Examples include:
- The terminal "e" in the suffix is optional but recommended.
- The traditional "irine" suffix can be used on unsaturated nitrogen 3-membered heterocycles.
- Because "etine" and "oline" were previously used for similar-sized nitrogen heterocycles, they cannot be used consistently as a suffix for 4- and 5-membered unsaturated heterocycles.
- Because of their established use for those substances or purposes, the use of oxine, azine, and silane for pyran, pyridine, and silacyclohexane, respectively, is prohibited (Ehrenhauser, 2015).

1.6. ISOMERISM AND STEREOCHEMISTRY

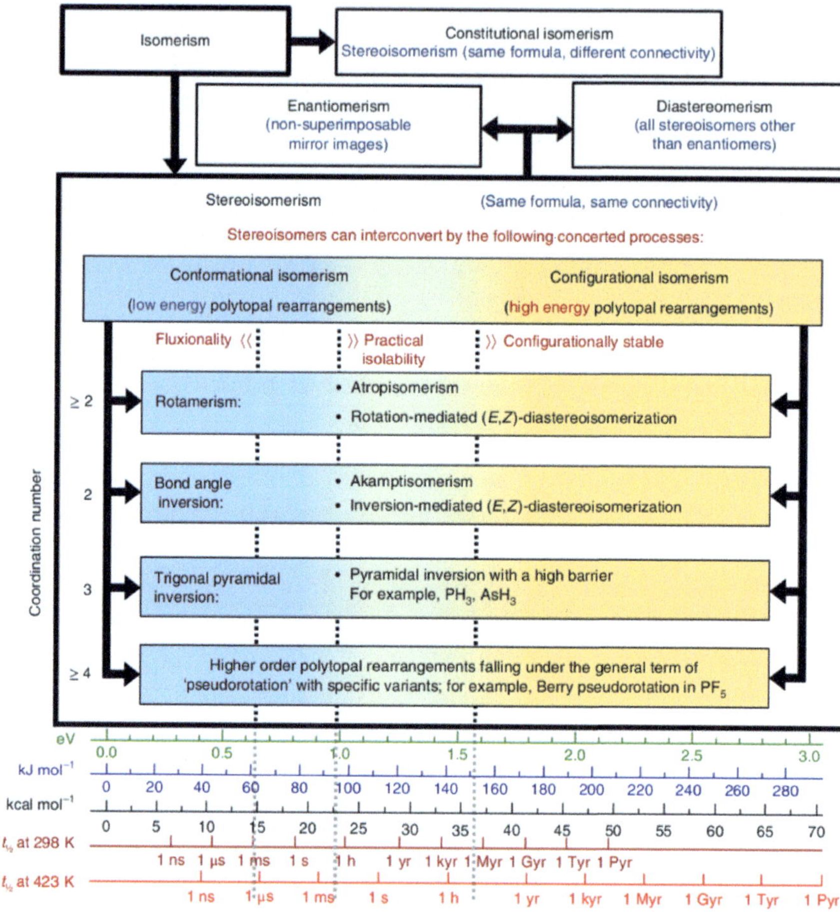

Figure 1.5. Isomerism hierarchy.

Source: *Canfield Peter, Blake Iain, Cai Zheng-Li, Luck Ian, Krausz Elmars, Kobayashi Rika, Reimers Jeffrey, & Crossley Maxwell, (2018); Scientific Figure on ResearchGate.*

Stereochemistry and isomerism are fundamental concepts in the IUPAC nomenclature system. The three-dimensional arrangement of atoms in a molecule is referred to as stereochemistry, whereas isomorphism refers to the phenomenon in which two or more compounds have the same molecular formula but different structural arrangements. To correctly name and describe chemical compounds, a solid understanding of stereochemistry and isomerism is required (Johnson, 2010) (Figure 1.5).

The two main types of isomerism are stereoisomerism and structural isomerism. Structural isomerism occurs when two compounds with the same molecular formula have different atom connectivity. This can occur due to different carbon skeleton arrangements, the presence of different functional groups, or the placement of substituents. Different physical and chemical properties differentiate structural isomers. Tautomeric isomerism, functional group isomerism, position isomerism, and chain isomerism are all subcategories of structural isomerism.

Stereoisomerism, on the other hand, is caused by the various spatial arrangements of atoms within a molecule. This is due to the presence of double bonds, chiral centers, or other asymmetries in the molecule. Geometrical (cis-trans) and optical (enantiomeric) stereoisomerism are two additional classifications for stereoisomerism (Ehrenhauser, 2015).

Geometric isomerism occurs in compounds with limited rotation around a double bond or ring system. When two carbon atoms are joined by a double bond, two distinct groups can be positioned on the same side of the bond (cis configuration) or on the opposite side (trans-configuration). The IUPAC nomenclature system accounts for geometrical isomerism by designating the corresponding configurations with the prefixes cis- and trans-.

Enantiomerism, also known as optical isomerism, occurs in compounds containing chiral centers. A chiral center is an atom, typically carbon, which is bonded to four distinct substituents. Chiral molecules are formed by enantiomers or pairs of mirror images of an isomer. The only physical distinction between enantiomers is their response to plane-polarized light. They rotate the polarized light plane clockwise (dextrorotatory, represented by the symbol +) or counterclockwise (levorotatory, represented by the symbol -). The IUPAC nomenclature system uses the Cahn-Ingold-Prelog (CIP) priority rules to assign R and S configuration labels to chiral centers, providing a systematic way to describe the absolute configuration of enantiomers. In addition to these types of isomerism, the IUPAC nomenclature system addresses stereochemistry through the use of stereochemical descriptors. These descriptors allow for the indication of the relative spatial arrangement of substituents within a molecule. The most common descriptors are cis/trans, E/Z (for compounds with multiple double bonds), and R/S (for chiral centers). These adjectives are critical for accurately describing the three-dimensional structure of a compound. Isomers are named and classified according to the IUPAC rules and guidelines based on their structural and stereochemical properties. Here are some important guidelines:

1. ***Structural isomerism*** refers to molecules that have the same molecular formula but different atom connectivity. Among the many structural isomers for which IUPAC provides naming guidelines are chain isomers, functional group isomers, position isomers, tautomers, and ring-chain isomers (Johnson, 2010).
2. ***Stereoisomerism:*** Despite having the same molecular formula and connectivity, stereoisomers differ in how their atoms are arranged in three-dimensional space. Geometric (cis-trans) isomerism and optical isomerism (enantiomerism) are the two types of stereochemistry that are the focus of IUPAC guidelines.
3. ***Geometric (Cis-Trans) Isomerism:*** Geometric isomers can form in cyclic structures or with a restricted rotation around a double bond. The specific substituents and their spatial arrangement must adhere to the IUPAC naming conventions for geometric isomers. Geometric isomers are also known as cis and trans.
4. ***Optical Isomerism (Enantiomerism):*** Enantiomers are mirror images of one another that cannot be superimposed. They differ in the spatial arrangement of one or more chiral centers (asymmetric carbon atoms). The (+)/(-) or (D)/(L) systems are recommended by IUPAC to denote the optical activity of enantiomers, and the R/S system to assign the configuration of chiral centers.
5. ***Chiral Centers and Chiralality*:** The IUPAC defines chirality as a molecule's inability to be superimposed on its mirror image. To form chiral centers, atoms, typically carbon, are bonded to four different substituents. Stereoisomerism is caused by a molecule with one or more chiral centers.
6. ***Cahn-Ingold-Prelog (CIP) Priority Recommendations:*** The CIP priority rules are typically used to determine the configuration of chiral centers. These rules prioritize substituents based on the atomic number of the atoms directly bonded to the chiral center. The highest priority group is labeled "1," and the other substitutes are assigned priority numbers based on that designation.

It is important to remember that, while these guidelines provide a framework for naming and categorizing isomers and stereochemistry, IUPAC reserves the right to update and improve them as new chemistry insights emerge (Ehrenhauser, 2015).

These isomerism and stereochemistry principles are incorporated into the IUPAC nomenclature system to ensure that chemical compounds are named and described precisely. When naming compounds, it is critical to consider isomers to avoid confusion. Names are assigned to structural isomers based on their distinct connectivity and functional groups. Butane and 2-methylpropane are two C_4H_{10} isomers, for example. Stereochemical descriptors, on the other hand, are required for stereoisomers.

To demonstrate how isomerism and stereochemistry are used in the IUPAC nomenclature system, consider the substance 2-chlorobut-2-ene. The double bond and chlorine atom introduce geometric and structural isomerism. If the chlorine atom is joined to the second carbon atom of the chain and the hydrogen atoms are on the same side of the double bond, the compound is known as (Z)-2-chlorobut-2-ene. The compound is known as (E)-2-chlorobutane. -2If the hydrogen and chlorine atoms are on opposite sides of the double bond; the compound is called -ene. Geometric isomerism is represented by the prefixes E and Z.

Finally, stereochemistry and isomerism are critical components of the IUPAC nomenclature system. Both structural isomerism and stereoisomerism, which are divided into geometrical and optical isomerism, are types of isomerism. While stereoisomers have the same connectivity but differ in the spatial arrangement of atoms, structural isomers do not. The IUPAC nomenclature system, which includes stereochemical descriptors such as cis/trans, E/Z, and R/S, provides guidelines and conventions for naming and describing isomeric compounds accurately. Understanding isomerism and stereochemistry is essential for effective communication and comprehension in the field of chemistry (Johnson, 2010).

In conclusion, the IUPAC nomenclature system, which provides a standardized method for naming chemical compounds, is a valuable resource in the field of chemistry. Its systematic approach ensures precision and clarity in communication, enabling scientists all over the world to successfully comprehend and communicate knowledge about chemical substances.

The IUPAC nomenclature system is based on strict adherence to a set of precise regulations and rules that apply to various classes of compounds, including organic, inorganic, and organometallic compounds. The IUPAC reviews and updates these regulations on a regular basis to account for new discoveries and advancements in the industry.

Using the IUPAC nomenclature system, chemists can precisely name compounds based on their molecular structure, functional groups, and other relevant characteristics. This facilitates the identification and categorization of substances and provides important information about their properties, reactivity, and potential applications (Ehrenhauser, 2015).Furthermore, the IUPAC nomenclature system promotes global collaboration and harmonization in chemical research and education. It serves as a unifying language that transcends linguistic and geographic boundaries, allowing scientists from various nations and backgrounds to communicate and share their findings with ease. Overall, the IUPAC nomenclature system has greatly aided the development of chemistry as a science. Its standardized approach to naming compounds promotes clarity, accuracy, and international cooperation, ultimately advancing our understanding of the fascinating world of chemicals and their interactions (Johnson, 2010).

CHAPTER

2 Nomenclature of Alkanes and Cycloalkanes

Contents

2.1. INTRODUCTION

Organic compound nomenclature is essential for precisely describing their chemical structures. Being fundamental hydrocarbons, alkanes and cycloalkanes adhere to a set of naming rules that promote clarity in chemical reactions and ease inter-chemist communication. This extensive page offers a thorough examination of the nomenclature of alkanes and cycloalkanes, encompassing IUPAC guidelines, typical prefixes and suffixes, and different examples to aid comprehension (NguyenHuynh, 2018).

Chemists can precisely describe and discuss the structures of distinct molecules because of the nomenclature of organic compounds, which acts as a universal language. Alkanes and cycloalkanes are important fundamental hydrocarbons within the diverse range of organic molecules. Chemists can precisely identify and categorize these compounds because of their systematic nomenclature, which is defined by the criteria set forth by the International Union of Pure and Applied Chemistry (IUPAC).

Alkanes, sometimes referred to as saturated hydrocarbons, are made up primarily of hydrogen and carbon atoms that are joined by a single covalent link. They are the foundation of many organic molecules and are crucial to the study of materials science, fuels, and medications. Contrarily, cycloalkanes are a subgroup of alkanes defined by the presence of carbon atoms organized in a closed ring or cyclic form. Cycloalkanes are characterized by their special arrangement, which also makes them common in both natural and manufactured chemicals (Pfefferle, 2018)

In order to successfully communicate, evaluate structures, and forecast chemical reactions, scientists must be familiar with the nomenclature of alkanes and cycloalkanes. The IUPAC system offers a standardized set of guidelines for naming these substances, guaranteeing uniformity and clarity among scientists. Chemists can accurately communicate the structure, connectivity, and presence of any functional groups or substituents in a molecule by adhering to these rules.

2.2. OVERVIEW OF ALKANES

Alkanes are a class of hydrocarbon compounds in which the carbon and hydrogen atoms are connected by a single covalent link. They go by the name paraffin as well. They are the essential building blocks of organic chemistry and are the most prevalent and readily recognizable organic molecules in nature. Because they completely saturate the carbon skeleton and have the most hydrogen atoms connected to each carbon atom, alkanes are known as saturated hydrocarbons.

Alkanes typically have the chemical formula CnH_{2n+2}, where "n" refers to the molecule's number of carbon atoms. According to this equation, the proportion of carbon and hydrogen atoms in alkanes increases linearly with each subsequent member in the series. For instance, the most basic alkane is methane (CH_4), which is then followed by ethane (C_2H_6), propane (C_3H_8), butane (C_4H_{10}), and so forth. Alkanes have different physical and chemical characteristics depending on the size, shape, and intermolecular forces of their molecules (Figure 2.1).

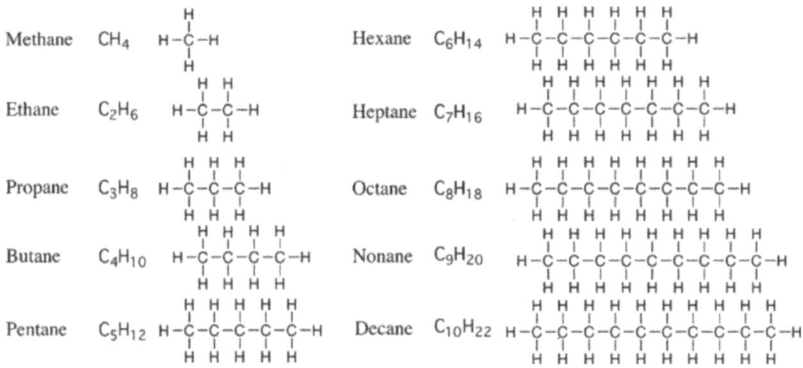

Figure 2.1. Alkanes, composition, constitution, and configuration.

Source: *Vančik, (2022); Springer, Cham.*

***Physical Characteristics*:** Alkanes are typically nonpolar, colorless, and odorless substances. Alkanes have minimal solubility in polar solvents due to their nonpolar nature but are soluble in nonpolar solvents like hexane or petroleum ether. Due to higher London dispersion forces between molecules, alkanes' melting and boiling temperatures rise as their carbon atom count rises. At room temperature, alkanes with shorter carbon chains are usually gases, but those with longer chains are liquid or solid (NguyenHuynh, 2018).

***Chemical Properties*:** Alkanes are rather stable substances under normal circumstances, according to their chemical properties. Due to the molecule's fully saturated state and the strength of the carbon-carbon single bonds, they are stable. The majority of acids, bases, and oxidizing agents are largely inert to alkanes. However, in the presence of oxygen, they can go through combustion reactions that result in the production of carbon dioxide and water as well as the release of heat energy (Pfefferle, 2018).

Reactions and Functionalization: Alkanes' reactivity is typically increased through functionalization, which involves replacing one or more hydrogen atoms with functional groups. These functional groups can be introduced through a variety of processes, including halogenation, hydrogenation, and oxidation. Alkyl compounds, also known as functionalized alkanes, are essential constituents of a wide range of organic and inorganic chemicals as well as intermediates in a number of chemical processes.

Applications in Industry and Daily Life: Alkanes are used in many different sectors. They are essential to the manufacturing of fuels including aviation fuel, diesel, and gasoline. Alkanes and its derivatives are utilized in the production of plastics, synthetic fibers, medicines, and agrochemicals as solvents, lubricants, and raw materials. The world's energy business heavily depends on the production and processing of petroleum, which is full of alkanes.

Alkanes are the cornerstone of organic chemistry and are essential to many disciplines of science and industry. Their straightforward composition, stability, and reactivity serve as a foundation for comprehending more complicated chemical molecules. Chemists can learn more about the fundamentals of organic chemistry and how to work with carbon-based compounds to create novel materials, drugs, and technology by researching alkanes.

2.3. OVERVIEW OF CYCLOALKANES

A closed ring or cyclic structure of carbon atoms distinguishes cycloalkanes, a subclass of alkanes. Cycloalkanes, like alkanes, are made up entirely of carbon and hydrogen atoms that are joined by a single covalent link. In contrast to their linear counterparts, cycloalkanes have special characteristics and reactivity due to the cyclic arrangement of carbon atoms.

Depending on the size of the ring and the kind of substituents connected to the ring, cycloalkanes display a variety of physical and chemical characteristics. Cyclopropane (C_3H_6), which has a three-carbon ring, is the most basic cycloalkane.

It is followed by cyclobutane (C_4H_8), cyclopentane (C_5H_{10}), cyclohexane (C_6H_{12}), and so on. Cycloalkanes that are bigger and have more carbon atoms in the ring adopt more stable conformations as they become less strained (Pfefferle, 2018).

Physical Characteristics: In contrast to their linear cousins of comparable molecular weight, cycloalkanes are often colorless and have a lower boiling point. This is mainly because cycloalkanes lack terminal methyl groups, which limits the amount of surface area open to intermolecular interactions. Since larger rings enable more efficient packing of molecules in the solid state, cycloalkane melting temperatures likewise rise with ring size.

Chemical Properties: Due to the high binding energies of carbon-carbon single bonds and the hydrogen saturating carbon atoms, cycloalkanes are relatively inert molecules. Under the right circumstances, people can have specific reactions. Acyclic (open-chain) alkanes can develop as a result of ring-opening reactions. In substitution processes, a functional group takes the place of a hydrogen atom on the ring in cycloalkanes. Ring strain and the presence of substituents are two examples of variables that may have an impact on the reactivity of cycloalkanes.

Conformational Isomerism: Cycloalkanes display conformational isomerism, which results from the ring's capacity to take on several three-dimensional conformations. The chair conformation, which reduces ring strain and increases the beneficial interactions between hydrogen atoms, is the most stable conformation for cyclohexane. Through ring-flipping processes, other less stable conformations, such as the boat and twist-boat conformations, can interconvert.

Applications: Cycloalkanes are used in a wide range of industries, such as materials research, medicines, and organic synthesis. Particularly cyclohexane is frequently utilized as a starting ingredient for the synthesis of nylon and other polymers as well as a solvent for a variety of chemical processes. In the synthesis of complex chemical molecules, functional groups or substituents linked to the ring of cycloalkanes act as important intermediates.

Due to their cyclic arrangement of carbon atoms, cycloalkanes form a significant class of organic molecules with unique structural and reactivity features. Their research offers insightful understanding of the fundamentals of organic chemistry and enables the design and synthesis of numerous organic compounds with various characteristics and functions. Chemists can increase their understanding of organic compounds and create novel applications in a variety of scientific and industrial fields by comprehending the distinct characteristics and behavior of cycloalkanes (Pfefferle, 2018).

2.4. INTERNATIONAL UNION OF PURE AND APPLIED CHEMISTRY (IUPAC) NOMENCLATURE FOR ALKANES

Standard naming conventions for organic compounds, including alkanes, are established by the IUPAC. To ensure clarity and accuracy in communication among chemists worldwide, the IUPAC nomenclature system offers a systematic and uniform approach to naming organic compounds. Chemists can give alkanes clear names by adhering to the IUPAC standards, which makes it easier to represent and comprehend their chemical structures.

The building blocks of organic chemistry are alkanes, which are hydrocarbons made of carbon and hydrogen atoms joined by a single covalent link. They are present in a wide variety of natural and artificial materials and serve as the structural core of many organic molecules. Alkanes are designated by the IUPAC nomenclature system, which is based on a set of guidelines that consider the chain's total number of carbon atoms, the existence of substituents or functional groups, and the relative importance of various substituents or functional groups. The main goal of IUPAC nomenclature is to give chemists a systematic and clear way to name alkanes so that they may communicate efficiently and convey accurate structural information. Chemists can recognize and distinguish between various isomers, conformations, and functionalized derivatives of alkanes by giving each chemical a distinctive name.

The IUPAC nomenclature for alkanes uses a hierarchical system in which the parent chain is identified as the longest continuous carbon chain and any substituents or functional groups linked to it are named. Chemists can visualize the compound's three-dimensional arrangement thanks to this methodical naming process, which guarantees that the compound's name appropriately reflects its structural properties. The IUPAC nomenclature for alkanes not only offers a uniform naming system, but also enables chemists to predict the physical and chemical properties of compounds based on their structural characteristics. It offers a framework for comprehending the connection between alkanes' structure and reactivity, making it easier to design and synthesize novel molecules with certain features and functions.

For chemists in a variety of domains, such as organic synthesis, drug development, materials research, and environmental chemistry, it is crucial to comprehend and use the IUPAC nomenclature for alkanes. Effective communication, confusion prevention, and the sharing of scientific knowledge all benefit from precise and uniform labeling (Pfefferle, 2018).

2.5. NAMING STRAIGHT-CHAIN ALKANES

The IUPAC has produced a set of guidelines for the systematic nomenclature of straight-chain alkanes. These rules ensure that the names of organic compounds are consistent and clear, enabling scientists to describe the structures of alkanes with clarity and effectiveness. The IUPAC nomenclature for straight-chain alkanes specifies the parent's name, the parent number of carbon atoms, and any substituents or functional groups that are present.

1. *Identify the Carbon Chain's Length*: The longest continuous chain of carbon atoms must be counted in order to determine the name of a straight-chain alkane. For naming purposes, this chain acts as the parent chain.
2. *Name the Parent Chain*: Using the following prefixes, the name of the parent chain is determined from the number of carbon atoms in the chain:

- For each carbon atom, use meth.
- Eth- for two atoms of carbon
- Three carbon atoms as a prop
- The four carbon atoms, however,
- Five carbon atoms in pent-
- Hex- for six atoms of carbon
- Seven carbon atoms in hept-
- For eight carbon atoms, oct-
- For nine carbon atoms, non-
- For 10 carbon atoms, dec-

3. *Indicate the Bond Type*: The suffix "-ane" is appended to the parent chain name since alkanes only have one covalent bond.
4. *Combining the Parent Chain Name and Suffix*: The straight-chain alkane's name is formed by combining the parent chain name and the suffix "-ane." A six-carbon chain, for instance, would be called "hexane."
5. *Indicate the Position of Substituents*: If the parent chain has any substituents (groups or atoms), their positions are denoted by numbers. The chain end closest to a substituent is where the numbering begins. When there are several substitutes, the parent's name is listed first followed by the substitutes in alphabetical order. Commas are used to separate the position numbers from

one another, and a hyphen is used to separate the position number from the substituent name.

6. *Examples*: The chemical is known as 2-methylbutane if a methyl group (-CH_3) is added to the second carbon atom of a butane chain, which has four carbon atoms.

The molecule is known as 3,4-dimethylhexane if it contains two methyl groups connected to the third and fourth carbon atoms of a hexane chain (six carbon atoms).

These guidelines enable scientists to name straight-chain alkanes in a systematic manner while guaranteeing clear and unambiguous communication of the compounds' structures. Understanding and competence in naming straight-chain alkanes in accordance with the IUPAC nomenclature system can be further improved with the use of practice problems and examples (NguyenHuynh, 2018).

2.6. NAMING BRANCHED ALKANES

The IUPAC has set criteria for the systematic nomenclature of branched alkanes. These rules ensure that the names of organic compounds are consistent and clear, enabling chemists to communicate clearly and accurately about the structures of branched or substituted alkanes (Pfefferle, 2018) (Figure 2.2).

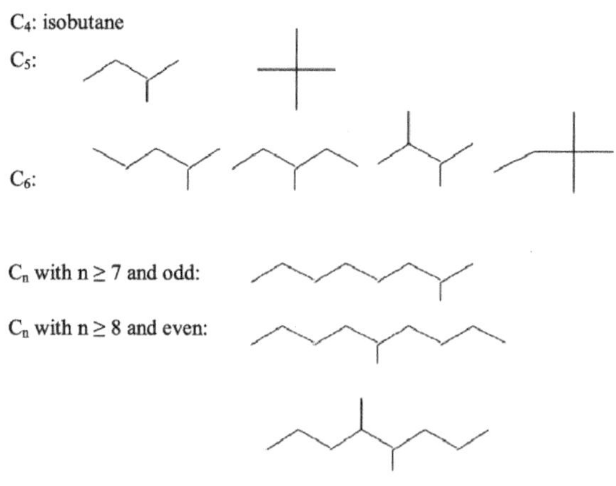

Figure 2.2. Model compounds for branched alkanes.

Source: *Burkle-Vitzthum Valerie, Bounaceur Roda, Marquaire Paul-Marie, Montel François, & Fusetti, (2011); Scientific Figure on ResearchGate.*

In the IUPAC nomenclature for branched alkanes, the longest continuous carbon chain is identified, a parent name is given, the carbon atoms are given a number, the positions of the substituents are indicated, and the substituents are given names.

1. *Finding the Longest Continuous Carbon Chain:* The longest continuous carbon chain, which acts as the parent chain for name purposes, is the first step in identifying a branched alkane. This chain may have branches or twists; it need not be a straight line.
2. *Assign the Parent Name:* The longest continuous chain of carbon atoms is used to determine the parent chain's name. To specify the number of carbon atoms, use the appropriate prefix (meth-, eth-, prop-, but-, pent-, hex-, hept-, oct-, non-, dec-).
3. Beginning at the end that is closest to the substituents, count the carbon atoms in the parent chain. Give the substitutes the lowest numbers you can, even if that means going the other way with the numbers (NguyenHuynh, 2018).
4. *Substituent Position:* To identify the position of substitutes on the parent chain, use the position numbers. Put the numbers before the substitute name, and use commas to separate groups of numbers. Use prefixes like di- (two), tri- (three), or tetra- (four) to denote the number of identical substituents if there are multiples of them.
5. *Name the Substituents:* Using the proper prefix or substituent name, identify the substituents that are a part of the parent chain. Methyl (CH_3-), ethyl (C_2H_5-), propyl (C_3H_7-), and butyl (C_4H_9-) are examples of common substituents.
6. *Combining the Names of the Parent Chain and Substituents:* The full name of the branched alkane is formed by combining the names of the parent chain and substituents. Before the parent chain name, alphabetize the substituents and hyphenate them to separate them. To maintain clarity, use any commas and hyphens that are required.
7. *Examples:* The molecule is known as 2-methyl-4-ethylpentane if it contains a methyl group (-CH_3) linked to the second carbon atom and an ethyl group (-C_2H_5) attached to the fourth carbon atom in a pentane chain (five carbon atoms).

The molecule is known as 2,5-dimethylhexane if it contains two methyl groups connected to the second and fifth carbon atoms of a hexane chain (six carbon atoms). These guidelines enable chemists to identify branched alkanes in a systematic manner while guaranteeing clear and unambiguous communication of the compounds' structures. Understanding and competence in naming branched alkanes in accordance with the IUPAC nomenclature system can be further improved with the use of practice problems and examples (Pfefferle, 2018).

2.7. NAMING ALKANES WITH FUNCTIONAL GROUPS

When alkanes have functional groups, which are specific atoms or groups of atoms that give the molecule certain chemical capabilities, additional naming conventions are utilized. The IUPAC provides guidelines for identifying alkanes with functional groups in order to provide systematic and understandable nomenclature. Following are the procedures to follow when naming alkanes with functional groups:

1. To determine the parent chain, find the longest continuous carbon chain that also contains the functional group. This chain will serve as the parent chain for naming reasons.
2. Assign the Parent Name: The parent chain is named using the appropriate prefix based on the number of carbon atoms, much like with straight-chain and branched alkanes.
3. The carbon atoms are counted: Beginning with the end that is closest to the functional group, count the carbon atoms in the parent chain. Even if it means numbering the functional group located in the opposite way, assign the lowest numbers you can to that position.
4. Determine and Identify the Functional Group: Determine the molecule's functional group. The parent chain name must be modified with the appropriate special suffix or prefix for Nomenclature for common functional groups includes:

- *Alcohols (-OH):* Change the parent chain's "-ane" ending to "-ol." Methane, for instance, changes into methanol (NguyenHuynh, 2018).
- *Aldehydes (-CHO):* Change the parent chain's "-ane" ending to "-al." Pentane, for instance, changes into pentanal.
- *Ketones (-C=O):* Change the parent chain's "-ane" ending to "-one." Hexane, for instance, changes into Hexanone.

- *Carboxylic acids (-COOH):* Change the parent chain's "-ane" ending to "-oic acid." For example, butane changes into butanoic acid.
- *Esters (-COO-):* Before the parent chain, name the alkyl group linked to the oxygen as a separate word, and change the "-ane" ending to "-oate." Ethane, for instance, changes into ethyl propanoate. each functional group.

5. Indicate the Position of Substituents: If the molecule contains substituents, identify their locations on the parent chain by giving each one a number. To connect a number and its substitute name, use hyphens rather than spaces to separate several numbers. According to where they are in relation to the functional group, the substituents are numbered.
6. Combining the Names of the Parent Chain, Substituents, and Functional Group: To create the whole name of the alkane, combine the names of the parent chain, substituents, and functional group. Put the names of the substitutes before the name of the parent chain, and separate them with a hyphen. For clarification, commas and hyphens should be used where needed.
7. Examples:

- The molecule is known as 2-pentanol if the second carbon atom of a pentane chain (five carbon atoms) has a hydroxyl group (-OH) connected to it.
- The molecule is known as 2-butanone if there is a carbonyl group (>C=O) between the second and third carbon atoms in a butane chain (four carbon atoms).

Chemists can methodically identify alkanes with functional groups by adhering to these rules, assuring correct representation and efficient communication of the compounds' structures.

When working with more complicated chemicals, it is crucial to resort to the IUPAC regulations for certain functional groups or to comprehensive nomenclature tables.

Understanding and competence in naming alkanes with functional groups in accordance with the IUPAC nomenclature system can be further improved with the use of practice problems and examples (Pfefferle, 2018).

2.8. INTERNATIONAL UNION OF PURE AND APPLIED CHEMISTRY (IUPAC) NOMENCLATURE FOR CYCLOALKANES

A standardized system for naming organic compounds, including cycloalkanes, is offered by the IUPAC. Cycloalkanes are cyclic hydrocarbons, meaning they are made up of hydrogen atoms connected to closed rings of carbon atoms. Due to their presence in a variety of natural products, medications, and polymers, these substances play a significant role in organic chemistry. Assigning distinctive and clear names to these cyclic structures is made easier by the IUPAC nomenclature system for cycloalkanes, which promotes global communication and understanding among chemists.

Cycloalkanes are named according to strict guidelines established by IUPAC, which provide clarity and uniformity in the nomenclature. The IUPAC system's main objective is to offer names for cycloalkanes that accurately reflect their structural characteristics and enable chemists to see how they are arranged in three dimensions. Chemists can successfully communicate the size, shape, and substituent positions of cycloalkanes through their names by adhering to these principles.

Several essential components are included in the nomenclature of cycloalkanes: The first step in naming a cycloalkane is to determine the size of the ring. This is done by counting the number of carbon atoms that make up the ring. Cycloalkanes can have three rings or more; for example, cyclopropane has three carbons, cyclobutane has four, cyclopentane has five, and so on. Choose a parent name: The number of carbon atoms in the ring determines the cycloalkane's parent name. For instance, cyclohexane, which has six carbons, and cyclooctane, which has eight carbons, are both names for rings. Number Positions of Substituents or Functional Groups: The number positions of any substituents or functional groups connected to the cycloalkane ring are given. Starting from any appropriate position, the carbon atoms in the ring are numbered sequentially, and the positions of the substituents are denoted by these numbers. Multiple substitutes are listed alphabetically with the lowest numbers possible. Combining the Parent Name and Substituent Names: To create the compound's full name, combine the names of the parent cycloalkane and any substituents or functional groups. Put the substitute names in front of the parent's name, then use hyphens to separate them. Multiple substituents are divided by commas.

Cycloalkanes are named consistently and unambiguously thanks to the IUPAC nomenclature system. It makes it possible for chemists to recognize and classify various cyclic structures, isomers, and functionalized derivatives. Additionally, it enables the design and synthesis of novel compounds with certain properties by enabling scientists to predict the physical and chemical characteristics of cycloalkanes based on their structural traits (NguyenHuynh, 2018).

For chemists working in organic synthesis, pharmaceutical chemistry, materials science, and several other domains, it is crucial to comprehend and utilize the IUPAC nomenclature for cycloalkanes. In order to facilitate effective communication, eliminate ambiguity, and promote the sharing of scientific information and discoveries, names should be precise and standardized. Chemists can confidently navigate the world of cyclic compounds and precisely describe their structures by knowing the IUPAC nomenclature for cycloalkanes.

2.9. NAMING SIMPLE CYCLOALKANES

Hydrogen atoms are joined to closed rings of carbon atoms to form cyclic hydrocarbons known as cycloalkanes. The IUPAC developed detailed criteria for naming these chemicals in order to ensure systematic and unambiguous nomenclature. In this part, we'll look at the nomenclature of unaltered simple cycloalkanes (Pfefferle, 2018).

Simple cycloalkanes are named using the following procedures:

1. The ring contains how many carbon atoms? Determine the cycloalkane ring's carbon atom count. Cycloalkanes can have three or more rings, such as those seen in cyclopropane (three carbons), cyclobutane (four carbons), cyclopentane (five carbons), cyclohexane (six carbons), and others.
2. Assign the Parent Name: The cycloalkane's parent name is determined by the number of carbon atoms in its ring. The root name of the alkane should be used, followed by the cyclo-prefix that corresponds to the number of carbon atoms in the ring. For instance, cyclopentane, a cycloalkane with five carbons, and cyclohexane, a cycloalkane with six carbons, are both given names.
3. Examples:

   Simple cycloalkane examples include the following:

- Cyclopropane, a three-carbon cycloalkane
- Cyclobutane, a cycloalkane with four carbons

- Cyclopentane, a cycloalkane with five carbons
- Cyclohexane, a cycloalkane with six carbons
- Cycloheptane, a cycloalkane with seven carbons
- Cyclooctane, a cycloalkane with eight carbons

Chemists can accurately communicate and identify simple cycloalkanes by using these naming rules to give these molecules distinct and precise names. It is vital to remember that there are extra requirements for naming substituents and marking their positions when cycloalkanes have functional groups or substituents attached to the ring. Chemical professionals who work with cyclic compounds must be familiar with the nomenclature of simple cycloalkanes. It facilitates the identification and characterization of various cycloalkane isomers and enables the straightforward conveyance of structural information. Chemists can navigate the world of cycloalkanes and make contributions to a variety of domains, such as organic synthesis, materials science, and medicinal chemistry, by becoming adept at the naming standards (NguyenHuynh, 2018).

2.10. NAMING SUBSTITUTED CYCLOALKANES

Cycloalkane compounds called replaced cycloalkanes have extra atoms or groups attached to the carbon atoms of the cyclic ring. These compounds are named in accordance with the guidelines provided by the IUPAC to ensure consistency and orderly naming. In this section, the nomenclature of substituted cycloalkanes will be discussed. The substituents must be identified, the carbon atoms in the ring must be numbered, and the names must be combined to form the compound name.

The steps to take while identifying substituted cycloalkanes are as follows:

1. ***Name and Identify the Substituents:*** Initially, determine which substituents are connected to the cycloalkane ring. Atoms or groups joined to the ring's carbon atoms can serve as substitutes. Alkyl groups (such as methyl and ethyl), halogens (such as chlorine and bromine), and functional groups (such as hydroxyl and carbonyl) are examples of common substituents. The IUPAC guidelines for identifying organic compounds are followed while naming each substituent.
2. ***Number the Carbon Atoms in the Ring:*** Beginning at any convenient location, count the carbon atoms in the cycloalkane

ring. The aim is to give the substituents the lowest numbers feasible. Choose the numbering scheme that yields the lowest set of numbers for all substituents when there are numerous substituents.

3. ***Determine the Position of the Substituents:*** Use numbers to determine the locations of the substituents on the ring. Put the numbers in front of the replacement name, separated by hyphens. If there are multiples of an identical substituent, use prefixes like di- (two), tri- (three), or tetra- (four) to indicate how many of them there are. Alphabetical order is used when there are multiple distinct substitutes.

4. ***The names of the substitutes and the parent cycloalkane should be combined:*** The names of the parent cycloalkane and the substituents are combined to form the compound's full name. Using hyphens to separate the names of the substituents, list the parent cycloalkane name first. Commas are used to separate several substituents (Pfefferle, 2018).

5. ***Examples:***

   Here are a few illustrations of substituted cycloalkane names:

- 1-methylcyclopentane is the name for a cyclopentane ring with a methyl group (CH_3) linked to the first carbon atom.
- A cyclohexane ring with a chlorine atom (Cl) joined to the second carbon atom: 2-chlorocyclohexane.
- a cycloheptane ring with a hydroxyl group (OH) connected to the third carbon atom: 3-cycloheptanol.
- 1-methyl-4-ethylcyclobutane is a compound with a methyl group (CH_3) linked to the first carbon atom and an ethyl group (C_2H_5) attached to the fourth carbon atom.

Chemists can precisely communicate and identify substituted cycloalkanes by using these naming rules to name and characterize substituted cycloalkanes. It is important to remember that if the substituent is a functional group, additional rules for designating functional groups are applied. These rules are covered in separate sections on naming alkanes with functional groups. Understanding the nomenclature of substituted cycloalkanes is essential for chemists working with complex cyclic molecules. It enables the examination of their chemical and physical properties, helps to identify different isomers, and enables the unambiguous communication of structures.

By mastering the naming conventions, chemists can navigate the world of substituted cycloalkanes and contribute to a number of fields, including chemical synthesis, drug discovery, and materials research (NguyenHuynh, 2018).

2.11. NAMING CYCLOALKANES WITH FUNCTIONAL GROUPS

Functional groups, which are particular atoms or groups of atoms that give the molecule certain chemical properties, can also be found in cycloalkanes. Additional guidelines are used when naming cycloalkanes containing functional groups to guarantee systematic and clear nomenclature. To appropriately express and convey the structures of these compounds, researchers can use the rules for naming cycloalkanes with various functional groups provided by the IUPAC.

The steps to take while identifying cycloalkanes containing functional groups are as follows:

1. Choose the Parent Chain: Choose the carbon chain with the functional group from the cycloalkane ring that is the longest continuous carbon chain. For naming purposes, this chain will act as the parent chain.
2. Choose a parent name: According to the information in the preceding sections, identify the parent chain by using the proper cycloalkane prefix based on the number of carbon atoms in the ring. Cyclohexane, for instance, is a cycloalkane with six carbons.
3. The carbon atoms of the cycloalkane ring should be numbered, commencing at any suitable position. Even if it means numbering the functional group located in the opposite way, assign the lowest numbers you can to that position.
4. Determine and Identify the Functional Group: Determine the molecule's functional group. To the parent cycloalkane name, a unique suffix or prefix for each functional group must be added (Pfefferle, 2018).

Following are a few typical functional groupings and their nomenclature:

- Hydroxyl group (-OH): Change the parent cycloalkane's "-ane" ending to "-ol." Cyclohexane, for instance, changes into cyclohexanol.

- If the carbonyl group (>C=O) is near the end of the parent chain, the "-ane" ending should be changed to "-al." Cyclobutane, for instance, changes into cyclobutanal.
- Replace the "-ane" ending with "-one" if the carbonyl group is located within the parent chain. Cyclopentane, for instance, changes into cyclopentanone.

5. Show the Position of the Substituents: Assign numbers to the places of the cycloalkane's substituents on the ring if any are present. To connect a number and its substitute name, use hyphens rather than spaces to separate several numbers. According to where they are in relation to the functional group, the substituents are numbered.
6. Combine the functional group name, substitutes, and parent cycloalkane names: To create the compound's full name, combine the names of the parent cycloalkane, any potential substituents, and the functional group. Put the names of the substituents before the name of the parent cycloalkane, and use hyphens to separate them. For clarification, commas and hyphens should be used where needed.
7. Examples:

- The chemical is known as 2-cyclohexanol if the second carbon atom in a cyclohexane ring has a hydroxyl group (-OH) connected to it.
- Cyclopentanone is the name of the chemical if the cyclopentane ring has a carbonyl group (>C=O) at the end.

Chemists can systematically name functional group-containing cycloalkanes by adhering to these rules, which will ensure correct representation and clear communication of the compounds' structures. When working with more complicated chemicals, it is crucial to refer to the IUPAC regulations for certain functional groups or complete nomenclature tables. Understanding and competence in naming cycloalkanes with functional groups in accordance with the IUPAC nomenclature system can be further improved with the use of practice problems and examples.

2.12. COMMON PREFIXES AND SUFFIXES

The study of molecules predominantly made of carbon and hydrogen atoms, as well as additional elements like nitrogen, oxygen, and halogens, is the focus of the chemistry field known as organic chemistry. These substances are the building blocks of life and are essential to many disciplines, such as medicine, materials science, and environmental science. To name organic compounds, which adhere to a methodical process outlined by the IUPAC, is crucial for communication and understanding among chemists (NguyenHuynh, 2018).

Prefixes and suffixes are key components of organic compound naming. These are particular syllables or word fragments that are appended to a compound's base name to signify different structural or functional groups that the molecule possesses. Chemists can precisely characterize the composition, structure, and properties of organic molecules by comprehending and making use of these frequent prefixes and suffixes (Pfefferle, 2018).

Prefixes that are typical in organic chemistry:

- Meth-: Denotes a single carbon atom.
- Eth-: Indicates a two-carbon atom.
- Prop-: Denotes a unit of three carbons.
- But: A four-carbon unit is indicated.
- Pent-: Describes a unit with five carbons.
- Hex-: Stands for a unit of six carbons.
- Hept-: Indicates a unit with seven carbons.
- Oct-: A unit of eight carbons.
- Non-: Describes a unit with nine carbons.
- Denotes a ten-carbon unit (dec-).

Whether they are straight-chain alkanes or cyclic compounds like cycloalkanes, these prefixes are used to denote the parent carbon chain in organic molecules.

Common Organic Chemistry Suffixes:

- Alkanes are denoted by the suffix -ane, which represents a single carbon–carbon bond.
- Alkenes are denoted by the suffix -ene, which denotes a double bond between carbon atoms.
- Alkynes are indicated by the suffix -yne, which denotes a triple bond between carbon atoms.
- -ol: Indicates the presence of an alcohol by way of the hydroxyl group (-OH).

- Aldehydes are indicated by the suffix "-al," which denotes the presence of a carbonyl group (C=O), typically at the end of the carbon chain.
- One: Indicates a ketone by a carbonyl group (C=O), typically found inside the carbon chain.
- Carboxylic acid functional group (COOH) is present and is indicated by the term "oic acid."
- -amine: A sign that an amino group is present ($-NH_2$).
- -amide: Indicates the existence of a functional group with the name "amide" ($CONH_2$).

These prefixes are used to name and describe particular structural elements or functional groups in organic molecules. Chemists can precisely name and describe organic compounds by adding prefixes and suffixes, indicating the number of carbon atoms in the chain, the type of bonds present, and the attached functional groups. The consistent application of these prefixes and suffixes guarantees the uniformity, originality, and informational content of compound names among the scientific community. When working with substituted or branched compounds, it is crucial to keep in mind that compound names can get more complicated when extra prefixes and suffixes are inserted to signify the location and kind of substituents. Nevertheless, comprehending the nomenclature of organic compounds requires a solid understanding of the fundamental prefixes and suffixes. The various prevalent prefixes and suffixes used in organic chemistry will be thoroughly examined and discussed in this article. For the purpose of improving comprehension and competence in recognizing and identifying organic compounds, we will discuss their definitions, uses, and examples. Understanding these prefixes and suffixes will help us to navigate the organic chemistry world more skillfully, promoting creativity, communication, and research in the subject (NguyenHuynh, 2018).

2.13. COMMON ALKYL GROUP PREFIXES

Alkyl groups are branches or substituents in organic chemistry that are created by taking one hydrogen atom out of straight-chain alkanes. As they can appear as substituents on parent carbon chains or as a component of functional groups, alkyl groups play a crucial role in the naming and identification of organic compounds. It is crucial to comprehend the typical alkyl group prefixes in order to correctly name and describe the structures of organic compounds (Pfefferle, 2018).

The following are some of the most typical alkyl group prefixes:

1. The simplest alkyl group is the methyl group (CH_3-), which consists of a single carbon atom joined to three hydrogen atoms. It is frequently included as a substituent in numerous chemical compounds and is frequently denoted as -CH_3.
2. Ethyl- (C_2H_5-): The ethyl group is made up of two carbon atoms joined together by a single bond, one of which is joined to three hydrogen atoms. It is frequently present in different chemical compounds and is denoted as -C_2H_5.
3. Three carbon atoms make up the propyl group (C_3H_7-), which are connected to one another by single bonds. Three hydrogen atoms are linked to the final carbon atom. The symbol for it is -C_3H_7.
4. Four carbon atoms in a straight chain, each connected to the next by a single bond, make up the butyl group (C_4H_9-). Three hydrogen atoms are linked to the final carbon atom. The symbol for it is -C_4H_9.
5. Five carbon atoms in a straight chain, each connected to the next by a single bond, make up the pentyl group (C_5H_{11}-). Three hydrogen atoms are linked to the final carbon atom. It's written as -C_5H_{11}.
6. Hexyl- (C_6H_{13}-): The hexyl group is made up of six carbon atoms that are arranged in a straight chain and are connected by single bonds. Three hydrogen atoms are linked to the final carbon atom. The symbol for it is -C_6H_{13}.
7. Heptyl- (C_7H_{15}-): The heptyl group is made up of seven carbon atoms that are arranged in a straight chain and are connected by single bonds. Three hydrogen atoms are linked to the final carbon atom. The symbol for it is -C_7H_{15}.
8. Eight carbon atoms are arranged in a straight chain and joined together by single bonds to form the octyl group (C_8H_{17}-). Three hydrogen atoms are linked to the final carbon atom. The symbol for it is -C_8H_{17}.
9. Nonyl- (C_9H_{19}-): The nonyl group is made up of nine carbon atoms that are linked together in a straight line by single bonds. Three hydrogen atoms are linked to the final carbon atom. Its chemical formula is -C_9H_{19}.

10. Ten carbon atoms in a straight chain, each connected to the next by a single bond, make up the decyl group ($C_{10}H_{21}$-). Three hydrogen atoms are linked to the final carbon atom. Its chemical formula is -$C_{10}H_{21}$.

These prefixes for alkyl groups act as name and descriptor building blocks for organic compounds, especially when substituents are joined to parent carbon chains. Chemists are able to precisely depict the structures and characteristics of a broad variety of organic compounds by fusing these prefixes with the appropriate parent chain names and functional group suffixes. In order to denote the presence of several identical alkyl groups in a chemical, these alkyl group prefixes can also be modified by the addition of number prefixes like di- (two), tri- (three), tetra- (four), etc. Alkyl groups can also branch, leading to more complicated substituent names. Alkyl group prefixes improve communication, make it easier to identify compounds, and advance our understanding of organic chemistry as a whole (Pfefferle, 2018).

2.14. COMMON CYCLOALKYL GROUP PREFIXES

The carbon atoms in cyclic alkyl groups known as cycloalkyl groups have a closed ring structure. These groups are essential in organic chemistry because they act as substituents in cyclic compounds, giving the molecules certain characteristics and reactivity. When naming and describing organic compounds with cyclic structures, it is crucial to comprehend the typical cycloalkyl group prefixes.

Here are a few of the most typical prefixes for cycloalkyl groups:

1. Cyclopropyl-: A three-membered carbon ring makes up the cyclopropyl group. It can serve as a substituent in a number of chemical compounds and is denoted as -C_3H_5.
2. Cyclobutyl-: A four-membered carbon ring makes up the cyclobutyl group. It can act as a substituent in organic compounds and is denoted as -C_4H_7.
3. Cyclopentyl: A five-membered carbon ring makes up the cyclopentyl group. It can serve as a substituent in a variety of chemical compounds and is denoted as -C_5H_9.
4. One of the most prevalent cycloalkyl groups is cyclohexyl, which is made up of a six-membered carbon ring. It is frequently found in organic compounds and is denoted as -C_6H_{11}.

5. The cycloheptyl group is made up of a carbon ring with seven members. It can serve as a substituent in organic compounds and is denoted as $-C_7H_{13}$.
6. Cyclooctyl: An eight-membered carbon ring makes up the cyclooctyl group. It can serve as a substituent in a variety of chemical compounds and is denoted as $-C_8H_{15}$.
7. Nine-membered carbon rings make up the cyclononyl group, which begins with this prefix. It can act as a substituent in organic compounds and is denoted as $-C_9H_{17}$.
8. Cyclodecyl: A ten-membered carbon ring constitutes the cyclodecyl group. It can serve as a substituent in a variety of chemical compounds and is denoted as $-C_{10}H_{19}$.

Using these cycloalkyl group prefixes, compounds with cyclic structures with alkyl substituents can be named and described. Chemists can precisely depict the structures and characteristics of organic compounds having cyclic components by fusing these prefixes with the relevant parent chain names and functional group suffixes (NguyenHuynh, 2018).

In order to denote the existence of numerous identical cycloalkyl groups in a compound, cycloalkyl groups can also be modified by adding number prefixes like di- (two), tri- (three), tetra- (four), etc. Further substituting cycloalkyl groups can result in more complicated substituent names.

In the discipline of organic chemistry, cycloalkyl group prefixes are systematically used to facilitate communication and understanding. In order to accurately represent and identify organic compounds, these prefixes offer a standardized method for describing the existence and arrangement of cyclic structures.

In conclusion, it is crucial for those working in organic chemistry to comprehend the nomenclature of alkanes, cycloalkanes, and their functional groups. To ensure consistency and clarity in scientific communication, the IUPAC has developed a systematic approach to naming these molecules. Chemists can precisely define the structure, content, and functional groups found in organic compounds by using the typical prefixes and suffixes covered in this article (Pfefferle, 2018).

Based on the number of carbon atoms in the chain and the location of substituents, straight-chain and branched alkanes are given different names. Similar to this, the names of cycloalkanes depend on how many carbon atoms are present in the ring structure. Additional prefixes and locants are added to the names of substituted alkanes and cycloalkanes to denote the location and kind of the substituents. The specificity and descriptive power of compound names are further increased by the inclusion of functional groups in the nomenclature. According to IUPAC nomenclature, functional groups including alcohols, aldehydes, ketones, carboxylic acids, amines, and amides are identified by certain suffixes.

Chemists are able to precisely communicate the structures and characteristics of organic substances by learning the concepts and regulations of naming. The efficient and accurate naming made possible by the use of standardized prefixes and suffixes facilitates scientific investigation, analysis, and collaboration in a variety of fields, such as medicine, materials science, and environmental chemistry.

In conclusion, a solid framework for precisely naming and describing organic molecules is provided by the systematic nomenclature of alkanes, cycloalkanes, and functional groups. The advancement of research, development, and application of these molecules in many sectors and scientific pursuits depends on the ability to effectively communicate and comprehend within the field of organic chemistry (NguyenHuynh, 2018).

CHAPTER

3 Nomenclature of Alkenes and Alkynes

Contents

3.1: INTRODUCTION

According to the language employed by IUPAC, ethene and ethyne can also be referred to by their respective synonyms, ethylene, and acetylene. This is the case even though these two compounds are not identical. In order to produce names for higher alkenes and alkynes, the relevant affixes are added to the root name of the unbranched alkane that has the same number of carbons as the higher alkene or alkyne that is being called. These affixes are "-ene" for higher alkenes and "-yne" for higher alkynes. When naming the initial carbon atom that participates in numerous bonds, the number that is lowest in the chain is prepended to the beginning of its name. The chain is given a number that corresponds to the multiple bonds, and after that, an alphabetical list of the parent chain's substituents is shown, with the positions being indicated by numbers (Hintermann, 2010) (Figure 3.1).

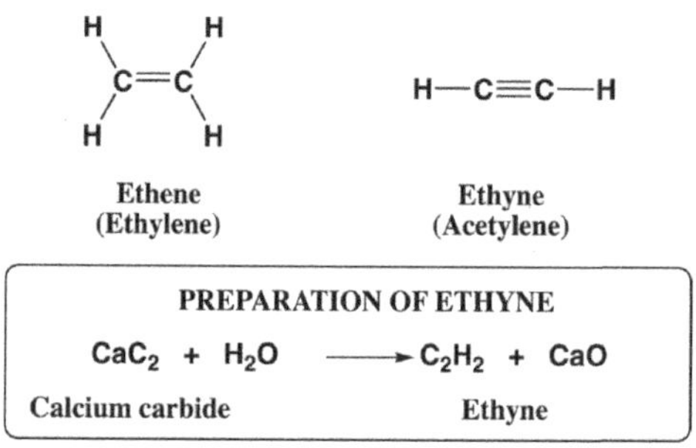

Figure 3.1. Alkenes and alkynes.

Source: *Vančik, (2022); Springer, Cham.*

Dienes are compounds that contain two double bonds, trienes have three, and so on down the line of an increasing number of double bonds. Dienes got their names from the addition of the suffix "-adiene" to the names of the alkanes to which they are structurally identical. The numbers 1 through 6 denote the sites of their double bonds and may be found in the locants. It is stated that the double bonds in dienes are cumulated if they create a C=C=C unit, a C=CC=C unit, or a C=C(CXY)nC=C unit. It is said that the double bonds in dienes are conjugated if they produce a C=CC=C unit, and it is said that they are isolated if they do not produce a C=CC=C unit.

The production of cycloalkenes requires just the addition of double bonds to rings of any size in order to be successful. In the process of assigning numerical values to the various substituted cycloalkene derivatives, the double bond is used as the point of departure in the counterclockwise direction.

In contrast to the rotation that takes place around single carbon-carbon bonds, which takes place at a reasonably quick pace, the rotation that takes place around double carbon-carbon bonds takes place only very infrequently. As a result of the fact that no two carbon atoms' substituents are the same, alkenes have the potential to give rise to stereoisomerism as a consequence of this feature. In order to distinguish between stereoisomerically different alkenes, the terms "cis" and "trans" are widely used as descriptors in chemical nomenclature. (An additional technique makes use of the CIP scheme and contains the prefixes E and Z.) Cycloalkenes with eight carbon atoms or more can exist as either the trans or cis stereoisomer. There is no way to separate trans-cycloalkenes that have less than seven carbons since it is not conceivable (Lippincott, 2018).

Because of the linear character of the CCCC unit that constitutes an alkyne, cycloalkynes are only conceivable when there is a significant amount of carbon present in the ring to enable room for the flexibility that is necessary to take this shape. This is the only circumstance in which cycloalkynes may exist. The cycloalkyne known as cyclooctyne has the formula C_8H_{12} and is the smallest and most stable of all the cycloalkynes that have been discovered (Hintermann, 2010).

Very little amounts of ethylene, which acts as a hormone in plants, are produced during natural processes. Ethylene promotes the growth of fruit by blocking the action of an enzyme that is responsible for the production of a new amino acid. This results in the fruit developing more quickly.

There is a possibility that essential oils extracted from a wide variety of plant species contain appreciable amounts of alkenes. (The distinctive scent, which is frequently referred to as the "essence" of a plant, is given to the client in the form of essential oils.) Myrcene is an alkene that can be found in bayberry, while limonene is an alkene that can be found in lime oil. Both of these alkenes are present in essential oils. In addition to its application as a paint solvent, pinene is a precursor that is required in the production of synthetic camphor, various medications, and other substances. Turpentine oil includes a substantial quantity of pinene, which is a hydrocarbon that is formed from the exudate of pine trees (Vančik, 2022).

Plant pigments such as lycopene, which is responsible for the characteristic red color of ripe tomatoes and watermelons, are examples of naturally occurring double-bonded hydrocarbons. Carotene and anthocyanin are two further examples of naturally occurring hydrocarbons that include double bonds. Polyene lycopene is a member of the carotenes family and is found in tomatoes. It possesses a total of 40 carbon atoms and a lot of double bonds throughout its structure (Lippincott, 2018).

Because of the way in which its single bonds and double bonds switch places with one another, lycopene is an example of a system that may be described as being conjugated. The degree to which unsaturated molecules are conjugated is directly proportional to the amount of light that may be absorbed by those molecules. Simple alkenes do not have a color that can be distinguished since they are transparent to ultraviolet (UV) light. When compared to saturated molecules, unsaturated ones are capable of absorbing more light at longer wavelengths. Due to the presence of conjugated double bonds in unsaturated compounds, this is the case. Polyenes that have extensive conjugation zones appear yellow to red in color because they have the ability to absorb visible light.

The hydrocarbon component of natural rubber, which makes up around 98% of the material, is made up of approximately 20,000 structural units per polymer molecule that are identified as C_5H_8. The molecules that go into the creation of this polymer are connected to one another in a very particular arrangement.

There are certain plants and fungi that have triple-carbon bonds however, these bonds are far rarer and less common than their double-carbon bond counterparts (Hintermann, 2010).

It is feasible to manufacture lower alkenes (up to four-carbon alkenes) at a cheap cost by isolating and dehydrogenating the hydrocarbons that are contained in natural gas and petroleum (for more details, see Alkanes: Chemical Methods). This may be accomplished by following the steps outlined in the previous sentence. More than 75 million metric tons of ethylene are produced across the world each year for use in industrial applications. On a yearly basis, the output of 1,3-butadiene (CH_2=CHCH=CH_2) is somewhere in the neighborhood of 2 MMT. Almost half of the ethylene that is produced on the planet is consumed in the process of making polyethylene. Polystyrene, vinyl chloride, and ethylene glycol antifreeze are some of the waste products generated during the polymerization process. Styrene, ethylene oxide, and vinyl chloride are the initial components in the polymerization process. This process also begins with styrene. The chemical propylene is the principal raw ingredient used in the production of polypropylene, and this manufacturing process is highly dependent on it. The production of synthetic rubber begins with the chemical compound known as 1,3-butadiene (Vančik, 2022). During the synthesis of higher alkenes and cycloalkenes, elimination techniques are utilized rather often. This is because these methods make it feasible to incorporate a double bond into a precursor that is already saturated, which is the reason for this result.

In a wide number of chemical reactions, including those involving alcohols and alkyl halides, the process of dehydration can take place. This is because dehydration is an end product of the reaction.

In most cases, the intellectual component of these strategies is stressed over their potential for monetary benefit. In the process of producing alkenes from alkynes by partial hydrogenation, alkynes create alkenes as a byproduct.

In order to produce acetylene and ethylene from hydrocarbons, one must first dehydrogenate the hydrocarbons and then separate the products of this process. However, the temperature at which acetylene is made is around 1,150 °C, but the temperature at which ethylene is generated is approximately 800 °C (1,500 °F). When compared to the need for ethylene, the quantity of acetylene, an industrial chemical, is in substantially lesser demand. Because ethylene is a more economical beginning material than acetylene, the majority of the compounds that are produced from acetylene can be synthesized more effectively from ethylene. This is because ethylene is a cheaper starting material. Either the addition of acetylene or the double elimination of a dihaloalkane, which is the process of eliminating both halogen atoms from a disubstituted alkane, can be used to produce higher alkynes. The addition of acetylene is the more common method of producing higher alkynes.

There is no similarity between the chemical properties of alkenes and alkynes and those of benzene or any of its derivatives. Alkenes and alkynes are wholly unique from benzene in this regard (Lippincott, 2018).

The structural structure of carbon compounds typically consists of double bonds, which is a property of the carbon atom. Numerous low-molecular-weight alkenes, like ethylene and propene, amongst a great number of others, have important economic functions in today's industrialized civilization. The alkenes in question are put to use in a wide range of other contexts as well. Ethylene is the organic compound that is produced in the greatest quantity, making it a central topic in the field of organic chemistry. Greater than 20 billion kg are produced each year in manufacturing in the United States. Ethylene is a very rare element due to the fact that it is only found in the natural world at extremely trace levels. When hydrocarbons are subjected to high temperatures, a process known as thermal cracking occurs, which leads to the generation of the massive amounts necessary to satisfy the demands of the chemical industry. The thermal cracking of the minute amounts of ethane that are recovered from natural gas is the process that generates the majority of the ethylene that is produced in the United States and other countries that have considerable natural gas reserves. This process is also used in other countries. Saturated hydrocarbons may be cracked at high temperatures to produce unsaturated hydrocarbons and hydrogen at the same time. This process is known as thermal cracking. Ethane can give rise to the synthesis of both ethylene and hydrogen if it is heated for a short period of time in a furnace at temperatures ranging from 800 to 900 degrees Fahrenheit. This process only takes a few minutes (Vančik, 2022).

The molecular formula consists of 2H Ethanol and 2H Ethylene. The production of ethylene in Europe, Japan, and other locations with limited natural gas supplies is virtually completely dependent on the thermal cracking of petroleum (Lippincott, 2018).

Because electron pairs in the valence shell have a tendency to be antagonistic toward one another, we predict that the bond angle that exists between the two carbon atoms that make up a double bond will be precisely 120 °. This is because electron pairs in the valence shell tend to be antagonistic against one another. The value of 121.7 ° for the angle of the HJCJC bond in ethylene that was predicted by VSEPR is an accurate representation of the true value. The angle of other alkenes may be off from the 120 ° that were predicted for it if there is stress between groups that are linked to either one or both of the double bond's carbons. One illustration of this would be the angle, which measures 124.7 ° and is present between the CJCJC bonds in propene (Hintermann, 2010).

Developing a Link between Space and Time Equivalent Carbon Dioxide Construct a model of the process that results in the production of a double bond between two carbon atoms in order to establish a connection between space and time equivalent carbon dioxide. A carbon-carbon double bond is composed of two individual bonds: one sigma and one pi. Two separate carbon-carbon single bonds combine to form a carbon-carbon double bond. In order to properly create a sigma bond with three other atoms, each carbon atom that is part of a double bond needs to make use of all three of the sp 2 hybrid orbitals that it possesses. The pi bond in a carbon-carbon double bond is oriented in a direction that is perpendicular to the plane created by the axes of the three sp 2 hybrid orbitals. This ensures that the double bond is stable. Pi bonds are made up of unpaired 2p orbitals in the carbon atoms, as the name suggests. It takes roughly 264 kJmol to spin one carbon by 90 degrees with respect to the other carbon such that the 2p orbitals on contiguous carbons are no longer able to overlap, which results in the breakage of the pi bond in ethylene. In order to do this, this must first be done. The amount of energy that is required to spin around a carbon-carbon double bond is noticeably more than the amount of thermal energy that is present in the molecule when it is at a temperature that is considered normal. It is possible to rotate around a single carbon-carbon bond, such as that found in ethane, just as it is possible to spin around a double carbon-carbon link, such as that found in ethylene.

This is because both types of bonds are made up of carbon atoms. Ethane has a carbon-carbon single bond, while ethylene has a carbon-carbon double bond. The carbon-carbon single bond in ethane is more easily rotated than the carbon-carbon double bond in ethylene. C isomerization of alkenes is the process that occurs when two different groups are connected to each carbon of a double bond in an alkene. This results in an altered structure of the alkene. The cis-trans isomerism process is another name for this phenomenon.

To have a clearer idea of what I mean, let's take a look at the molecule known as 2-butene. In contrast to cis-2-butene, which has two methyl groups placed on the same side of the double bond, trans-2-butene has both of its methyl groups situated on the opposite side of the bond. At room temperature, it is not possible to transform these two compounds into one another because their physical and chemical characteristics are different and they cannot be converted into one another.

This is because there is a finite amount of rotation around the double bond. It takes 264 kilojoules of energy to break the ethylene P bond, which is created when the 2p orbitals of the two carbons that are near to one other do not line up with each other. This occurs when two carbons are in close proximity to each other.

When a material is at the normal temperature that is considered to be ambient, there is not enough thermal energy available in the substance for the substance to spin around a carbon-carbon double bond. Space-filling simulations of the cis and trans isomers of 2-butene reveal that the cis isomer is less stable than the trans isomer. This is because the cis isomer has a non-bonded contact strain between the alkyl substituents that are on the same side of the double bond as the double bond in the cis isomer. The equatorial orientation of methylcyclohexane is preferred, despite the fact that both orientations endure the same total amount of steric strain (Vančik, 2022).

3.2. THE ALKYNE STRUCTURE IN THE CONFIGURATION

The existence of a carbon-carbon triple bond is what differentiates alkynes, which are considered to be functional groups. The simplest alkyne in terms of its molecular structure is ethyne, which has the chemical formula C_2H_2 and may be written as ethene. The molecule of ethyne is absolutely linear, and every angle that is generated by its bonds is precisely 180 degrees. According to the orbital overlap model, triple bonds are composed of a sigma bond, which is generated by the overlap of two sp hybrid orbitals of neighboring carbons, a pi bond, which is generated by the overlap of two parallel 2py orbitals, and a second pi bond, which is formed by the overlap of two parallel 2pz orbitals. An extra pi bond has been generated as a result of the overlap of two parallel 2pz orbitals, and this pi bond is what holds these three bonds together. When the sp hybrid orbital of each carbon in the ethene molecule coincides with the 1s atomic orbital of a hydrogen atom, this results in the formation of a covalent bond between the two atoms of carbon and hydrogen in the ethene molecule (Hintermann, 2010).

3.3. WHAT DOES IT MEAN WHEN THEY ARE REFERRED TO AS ALKENES AND ALKYNES?

As we are about to see in a minute, in addition to the IUPAC designations that are given to various alkenes, their common names are also used to refer to them. References from the IUPAC, in order to construct IUPAC names for alkenes, the -an prefix must first be removed from the name of the parent alkane, and then the -en prefix must be added in its place. In light of this, the compound with the formula CH_2CH_2 is referred to as ethene, whereas the one with the formula CH_3CHCH_2 is called propene. When the position of the double bond has an influence, it is possible to differentiate between the various isomers of higher alkenes by making use of a numbering system. This is the case when the effect is present. The carbon atoms that are a part of the longest carbon chain that has a double bond have been assigned the lowest numbers that may possibly be assigned to them. After then, the position of the double bond will be designated by using the number of the first carbon. The naming of branched or substituted alkenes with names that are phonetically similar to the names of alkanes is a widespread practice. The double bond, as well as any and all substituents, the carbon atoms, and the main chain, are all given labels (Lippincott, 2018).

Take into consideration the fact that the chain of 2-ethyl-3-methyl-1-pentene has a whopping six carbons altogether. However, the molecule in issue is known as disubstituted 1-pentene since pentane is the parent hydrocarbon and the longest chain that includes the carbon-carbon double bond has only five carbons in it. For this reason, the molecule in question is referred to as disubstituted 1-pentene. In order to generate IUPAC names for alkynes, one must first remove the -an prefix from the name of the parent alkane and then add the -yn suffix in its place. Ethyne is the name given to the chemical molecule HCCH, whereas propyne is the name given to the complex CH_3CCH. Both of these compounds belong to the category of hydrocarbons. According to the IUPAC system, which is still in use today, the phrase HC-CH can alternatively be referred to as ethyne. This is because ethyne is an alternate name for the compound. An alternative possibility is acetylene. On the other hand, "acetylene" is a term that is applied a great deal more frequently than any of these other designations. When counting the longest carbon chain, getting the lowest feasible set of numbers for bigger compounds that comprise carbons that are triply bound is easiest to achieve by using the triple bond as both the starting point and the ending point for the counting process. We are able to pinpoint the location of the triple bond to an incredibly high degree of accuracy when we use the initial carbon number of the triple bond (Vančik, 2022).

Alkenes and alkynes with the same number of carbon atoms as alkanes and cycloalkanes behave in a manner that is quite similar to that of alkanes and cycloalkanes when it comes to their physical characteristics. The attractive interactions that take place between molecules in an alkyne are substantially greater than those that take place in an alkane or alkene. This is because the electric field of an alkyne can be altered more readily than that of an alkane or alkene. Since this is the case, the boiling points of alkynes are significantly higher than those of alkanes and alkenes (Hintermann, 2010).

3.4. THE COMPOSITION OF THE MOLECULE'S STRUCTURE

In comparison to alkanes, alkenes are capable of performing a far greater number of different chemical reactions. In most situations, all that alkanes do is simply replace the double or triple bond, whereas alkenes and alkynes really add to the structure rather than merely replacing it. Hydrogenation is the process that may be used to convert alkenes to alkanes, and it is conceivable to do so. During this process, atomic hydrogen, denoted by the symbol H_2, is incorporated into a multiple-bond structure. Only in the presence of metal catalysts that have been extremely finely divided, such as nickel (Ni), platinum (Pt), palladium (Pd), or rhodium (Rh), can the reaction occur at a rate that is regarded as being acceptable. Nickel (Ni), platinum (Pt), palladium (Pd), and rhodium (Rh) are all examples of such catalysts.

Hydrogenation is a process that is used to make alkanes and cycloalkanes, which are subsequently utilized to prolong the shelf life of highly unsaturated vegetable oils. These compounds are produced as a byproduct of the hydrogenation process. Both of these are the results of carrying out the hydrogenation procedure. Lipids that were formerly liquid can be converted into solid forms using these techniques, which are used on vegetable oils. The partial hydrogenation of soybean oil is the procedure that is required for the production of margarine and other butter substitutes (Lippincott, 2018).

Over the past several years, there have been tremendous advancements achieved in the investigation and development of enantioselective hydrogenation catalysts. Whenever chiral substances are subjected to enantioselective hydrogenation, there is a clear preference for either one of the two enantiomers (also known as potential structural forms). When developing enantioselective hydrogenation catalysts, it is common practice to use enantiomerically homogeneous ligands coupled to rhodium as the fundamental building block. Because of this, one of the carbon atoms that make up the double bond is converted into a stereogenic center. Enantioselectivities of higher than 90% of a single enantiomer are usual in enantioselective hydrogenations, which are often employed in the production of enantiomerically pure pharmaceuticals.

The reaction that takes place when alkenes are exposed to bromine and chlorine results in the formation of dihaloalkanes. The quick addition procedure may be carried out at room temperature and does not call for the use of a catalyst in order to be fruitful. When chlorine is combined with ethylene, a chemical with the name 1,2-dichloroethane is generated as an intermediate substance. One of the most important and widespread applications for ethylene chloride is in the production of vinyl chloride, and this chemical is one of the key ingredients in that process (Vančik, 2022).

Alkenes are reactive molecules that can be altered by adding compounds of the type HX, where X is a halogen or another electronegative group; the hydrogen atom of the compound forms a covalent bond with one of the carbon atoms in the C=C unit, and the X atom forms a covalent bond with the other carbon atom in the C=C unit. Alkenes are molecules that can be altered by adding compounds of the type HX. The addition of compounds of type HX can result in the formation of alkenes (Figure 3.2).

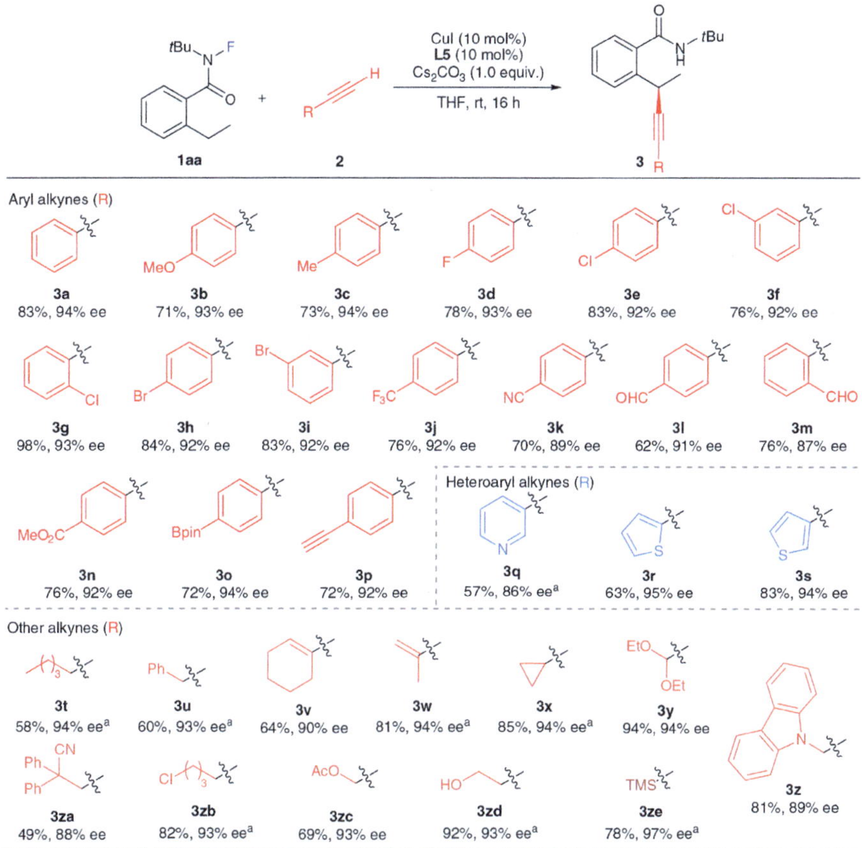

Figure 3.2. Substrate scope of alkynes standard conditions.

Source: *Zhang Zhen-Hua, Dong Xiao-Yang, Du Xuan-Yi, Gu Qiang-Shuai, Li Zhong-Liang, & Liu Xin-Yuan, (2019); Scientific Figure on ResearchGate.*

When HX is a strong acid, such as hydrochloric or hydrobromic acid, the reaction takes place fast and does not require the inclusion of an acid catalyst. This is because these types of acids are themselves acid catalysts. The reaction of ethylene with water in the presence of phosphoric acid is one of the ways that industrial ethanol may be made. Phosphoric acid is also required for the reaction (Lippincott, 2018).

When there is an imbalance in the number of hydrogen atoms that are directly bonded to the two carbon atoms that make up a double bond, the H in the compound HX is added to the carbon atom that has a greater number of directly attached hydrogen atoms, and the X is added to the carbon atom that has a lower number of directly bonded hydrogen atoms. This causes the number of hydrogen atoms that are directly bonded to the carbon atoms to become more balanced. Instead of n-propyl hydrogen sulfate ($CH_3CH_2CH_2OSO_3H$), which is created when propylene is exposed to sulfuric acid (H_2SO_4), isopropyl hydrogen sulfate ($CH_3CH_2CH_2OSO_3H$) is the product that is generated. This is due to the fact that the chain length of isopropyl hydrogen sulfate is much shorter than that of n-propyl hydrogen sulfate. The production of isopropyl alcohol using commercial means begins with the heating of isopropyl hydrogen sulfate in water. This is the first stage in the process. This stage takes place in a commercial or manufacturing environment (Hintermann, 2010).

It is claimed that a reaction is "regioselective" when it favors one route over another, such as when sulfuric acid is added to propylene. One can speak of a reaction as being regiospecific when it exhibits regioselectivity to an absolute degree. The Markovnikov rule is applied in order to make a prediction regarding the regioselectivity that will take place as a result of the addition of an unsymmetrical reagent such as HX to an unsymmetrical alkene such as $H_2C{=}CHR$ (Vančik, 2022).

Compounds with the formula R_2BH, such as boron hydrides, hydroborate, or produce organoboranes by reacting with alkenes. In the following step, these organoboranes might be oxidized using hydrogen peroxide (H_2O_2) as the oxidant. This would result in the production of alcohol. The final result is the same as it would have been if both hydrogen and oxygen had participated in the production of the double bond in the regioselective way suggested by the Markovnikov rule. The end result is the same as it would have been if both hydrogen and oxygen had participated in the formation of the double bond. The hydroboration-oxidation sequence is one of the several boron-based synthesis methods that were developed by Herbert C. Brown, a chemist who worked in the United States. Brown was a trailblazer in the industry he worked in (Lippincott, 2018).

When alkenes are subjected to a reaction with oxidizing agents such as potassium permanganate ($KMnO_4$) and osmium tetroxide (OsO_4), the product of this reaction is a class of compounds known as vicinal diols. These are some examples of compounds that contain OH groups attached to carbon atoms that are next to one another. The most common techniques involve the utilization of tert-butyl hydroperoxide ($(CH_3)_3COOH$) and catalytic quantities of osmium dioxide (OsO_4). These steps are carried out when oxidizing agents are present.

Epoxides are often derived from alkenes, and their structure is that of a three-membered ring that is composed of one oxygen atom, two carbon atoms, and one oxygen atom. Oxirane, which is also known as ethylene oxide, is produced by heating a mixture of ethylene and air (or oxygen) over a silver catalyst. This process results in the formation of ethylene oxide. The production of ethylene oxide, which may also be referred to by its other name, oxirane, is the end consequence of this process. It is generally agreed that oxirane is the most fundamental of all epoxides. Epoxides are involved in a diverse range of functions, acting as beneficial mediators in several of those processes. Ethylene oxide is initially converted into ethylene glycol, which is the major component of both automotive antifreeze and polyester fibers and films, for example. Ethylene oxide is also the primary component of antifreeze. In the lab, it is common practice to engage the reaction of an alkene with peroxy acid while attempting to synthesize an epoxide. This is one of the steps in the process (Vančik, 2022).

The Diels-Alder cycloaddition reaction involving conjugated dienes is an innovative method that has many important applications in the real world. In this procedure, an alkene and a conjugated diene go through a reaction that ultimately leads to the production of a molecule that has a cyclohexene ring as its centerpiece. The Diels-Alder cycloaddition is a particularly remarkable chemical process due to the fact that it is able to produce two carbon-carbon bonds even when no catalysts are used in the process. Because of its unique characteristic, it stands out among other chemical reactions. Otto Diels and Kurt Alder were awarded the Nobel Prize in Chemistry in 1950 for their discovery of this reaction and their demonstration that it may be employed in the manufacturing of synthetic materials. Both of these accomplishments led to the awarding of the Nobel Prize.

The generation of alkynes is a byproduct that can occur as a result of a variety of processes that include the use of alkenes. It is possible for the hydrogenation of alkynes to result in the creation of alkenes or alkanes. The creation of an alkane is the outcome of a process that involves two H_2 molecules and a triple bond and typically takes place under circumstances that are typical for catalytic hydrogenation.

This method may be utilized to successfully complete the synthesis of alkenes. Researchers have discovered new catalysts that are less active and can be poisoned in order to halt the reaction before it reaches the alkene stage. This will allow the process to be stopped. When both the trans and cis stereoisomers of an alkene are utilized as reactants in a reaction, the product of the reaction is always the cis stereoisomer of the alkene (Hintermann, 2010).

The reaction that takes place when an alkyne is coupled with either bromine or chlorine results in the formation of dihaloalkenes and tetrahaloalkanes as the products of the reaction.

Reactions are able to take place between electronegative atoms or groups and alkynes when the HX type of molecule is involved. The reaction between acetylene (HCCH) and HCN results in the production of acrylonitrile (CH_2=CHCN), whereas the reaction between acetylene (HCCH) and HCl results in the production of vinyl chloride (CH_2=CHCl). Hydrogen chloride is the substance that acts as a catalyst for each of these processes. Vinyl chloride and acrylonitrile are exciting starting materials for the synthesis of useful polymers because they can be synthesized more cheaply from an alkene. More specifically, vinyl chloride can be made from ethylene, while acrylonitrile can be created from propylene. On the other hand, neither of these compounds can be produced in appreciable quantities by synthesizing acetylene.

In order to prevent the formation of an unstable enol under the conditions of synthesis, the first product of alkyn hydration, which is an enol, is converted to an isomer that includes a carbonyl group. This is one of the factors that contribute to the extremely rare occurrence of the alkyn hydration reaction.

Acetylene and terminal alkynes have a substantially greater acidity level despite both being regarded as weak acids. This is in comparison to alkenes and alkanes, which are both considered to be weak acids. Using a strong base like sodium amide ($NaNH_2$) that has been dissolved in liquid ammonia, it is possible to remove triple-bonded hydrogens that are found in carbon (Vančik, 2022).

The great majority of situations call for adding sodium salt of the alkyne that was generated in this method to an alkyl halide in a concise amount of time. When you use this technique, you will end up with C-C bonds, which are necessary for the creation of higher alkynes.

The International Union of Pure and Applied Chemistry (IUPAC) devised a naming system for the class of triple-bonded hydrocarbons known as alkynes. Alkynes were given their names in accordance with this system. The following are some general guidelines that should be followed while naming alkynes. Determine which of the lengthy carbon chains with three bonds is the longest. This sequence accounts for a significant portion of the designation.

Counting the amount of carbon atoms requires beginning the process from the end of the chain that is closest to the triple bond. In order to accurately depict the triple bond, the lowest number feasible should be utilized. In the case that there is more than one triple bond, the choice of numbers should be made based on the lowest feasible total (Lippincott, 2018).

Make sure you use the right prefix when referring to the number of carbon atoms in the chain. The following prefixes are utilized when referring to the complete amount of carbon atoms present in the chain:

It takes two carbons to make meth, three carbons to make eth, four carbons to make prop, five carbons to make pent, six carbons to make hex, seven carbons to make hept, eight carbons to make oct, nine carbons to make non, and ten carbons to make Dec. For the remainder of it, please see above.

Simply adding "-yne" to the end of a compound's name will reveal that it is an alkyne.

To determine the precise position of the triple bond, it is necessary to identify the atomic number of the carbon atom that functions as the bond's first point of connection. Between the prefix and the number, you should put a hyphen (Hintermann, 2010).

In order to create a number, it is necessary to take into account the position of each substituent group along the carbon chain. When referring to substituent groups, use prefixes such as methyl-, ethyl-, and so on.

In alphabetical sequence, the names of any substitute groups that are used should come before the primary name. Use commas to denote the separation of several modifications.

As an illustration, let us have a look at some specific names for alkynes:

Beginning at the second carbon atom in the main chain of $CHCCH_2CH_3$, which is made of four carbon atoms, the triple bond in the $CHCCH_2CH_3$ molecule may be found. The IUPAC name for this substance is "2-butyne," and it is a chemical compound (Vančik, 2022).

In the molecule that may be represented by the formula $HCCCH_2CH_2CH_3$, there is a triple bond present between the first and second carbon atoms. You might be more familiar with it if I were to refer to it as 1-pentyne. The chemical formula for the triple bond is $CH_3CCCH_2CH(CH_3)_2$, and it is located on the second carbon atom of the main chain. The triple bond is called cyclohexane. The molecule's fourth carbon atom is connected to two different methyl groups by covalent bonds. 2-Methyl-4,6-hexadiyne is the term given to the molecule at issue when it is referred to by its chemical name.

It is essential to bear in mind that the aforementioned are only basic recommendations; specific constraints or changes may be required for enterprises that are more sophisticated. Examining the most recent version of the IUPAC nomenclature rules is an absolute must if one wants to achieve the highest level of accuracy possible.

If a molecule includes more than two triple bonds, the positions of those triple bonds should be described by inserting a prefix to the number. For example, "di-" should be appended to the number to indicate two triple bonds, "tri-" should be appended to indicate three triple bonds, and so on. Please provide an explanation as to the possible locations inside the fundamental structure of the chain where each triple bond can be located.

It is vital to begin the process of looking for alkyne groups inside a molecule by first establishing whether or not the molecule contains any functional groups. In order to refer to the functional group as a substituent, its name will be modified to add the prefix "yn-" after the alkyne. This will allow the group to be called a substituent. If a molecule, for instance, included both an alcohol group and an alkyne group, the alcohol portion of the molecule would be denoted by the letter "ol," and the alkyne would be denoted by the letter "yn-" as a substituent. In the event that the molecule had both kinds of groups, then this would be the situation (Lippincott, 2018).

Find the triple bond of the cyclic alkynes using the IUPAC criteria for cyclic compounds, and after that, determine the ring structure of the cyclic alkynes. Include not just the atomic coordinates of the triple bond but also the total amount of carbon atoms that are situated within the ring in your response.

When the carbon chain has a significant number of distinct substituents, it is important to give the lowest possible numbers based on the alphabetical order of the substituents. In the case that you have the opportunity to do so, you should start by giving the triple bond the smallest number feasible, and then proceed to give that number to each substituent in turn (Vančik, 2022).

When there are many occurrences of the same object, it is helpful to use prefixes such as "di-" for two, "tri-" for three, etc. It is important to make a mental note of where each substituent falls in the main chain. If many functional groups or substituents have the same alphabetical precedence, the one that occurs first in the name ought to be assigned the lowest possible number. This is because it has the most significance in the context of the name. This rule is applicable even if the priority is held jointly by a number of different functional groups or substituents.

When you put anything in brackets, it indicates that you are placing it in a new category. This is of the utmost significance in circumstances in which a great number of alternatives that are not significantly different from one another are available, or in which those concerned are distributed over a vast geographic region (Hintermann, 2010).

It is required to identify any breaks in an alkyne chain as substituents and give them the appropriate prefixes, such as "methyl-," "ethyl-," "propyl-," and so on. If this is not done, the chain cannot be properly analyzed. The assignment of a number to the branch allows for the location of that branch in reference to the main stem to be identified.

If there are a number of distinct approaches to an issue, organize the potential answers in alphabetical order, skipping any prefixes that may be associated with them. To mark the separation of several alterations, commas should be used.

A prefix such as "di-" or "tri-" is added to the name of a molecule in order to identify the number of alkyne groups that are included inside the molecule. This allows the number of alkyne groups to be easily identified. It would be helpful if you could present the alkyne groups in the main molecule in the sequence in which they are located in relation to one another (Lippincott, 2018).

If the alkyne comprises more than one functional group, the prefix "yn-" should be written before the name of the alkyne to indicate that it is a replacement. The functional group that is the most significant should be included first in the list of functional groups.

The suffix "-yne" is used in place of the suffix "-ol" as the final form of the name of the compound if an alkyne contains a functional group such as the alcohol (-OH) group. As a substitute for the alkyne, the prefix "yn-" should be added to the name of the alcohol in this circumstance so that it is clear that the alcohol is being used in its stead.

Some alkynes may nevertheless be recognized by their common names despite the fact that they do not conform to the criteria established by the IUPAC. Usually, names such as this one have some form of importance attached to them, whether it be historical or etymological in nature. For instance, ethyne, which has the formula C_2H_2, is sometimes referred to as acetylene, which also has the formula C_2H_2. Acetylene also has the formula C_2H_2 (Vančik, 2022).

Even while these standards ought to be acceptable for the overwhelming majority of alkyne identification scenarios, there is always the risk that certain circumstances will be more challenging. When one is working with one-of-a-kind substances or circumstances, it is often advised that one look at the most recent edition of the IUPAC guidelines or related specialized literature (Lippincott, 2018).

It is essential to bear in mind that the IUPAC nomenclature was created in order to give a means of identifying chemicals that is consistent as well as accurate, and it is for this reason that it was devised. In spite of the fact that the ideas discussed above account for the overwhelming majority of the situations in which alkynes have to be taken into consideration, it is possible that there are additional, more intricate cases as well. When dealing with unusual circumstances or chemicals, it is suggested that the most recent recommendations from IUPAC or specialized literature be researched. The vast majority of situations fit this description (Hintermann, 2010).

CHAPTER

4 Carboxylic Acids, Esters, Acyl Chlorides & Amides

Contents

4.1. INTRODUCTION

Nomenclature, which is used to name molecules precisely and consistently, is crucial to the study of organic chemistry. Nomenclature functions are a universal language that makes it possible to classify, identify, and communicate about organic substances effectively. We will examine the names of esters, acyl chlorides, acids, and amides in this document emphasizing their importance in organic chemistry.

For scientists to communicate effectively, accurate naming is crucial. By employing a consistent nomenclature system, organic chemists from various fields and nations can comprehend and replicate experiments, syntheses, and analyzes. It enables scientists to collaborate more effectively, share knowledge, and expand upon previous understanding (Nishihara, 2020).

Furthermore, the identification of organic compounds depends heavily on proper nomenclature. Chemists can determine the functional groups, molecular structures and substituents of esters, amides, carboxylic acids, and acyl chlorides by methodically identifying these compounds. This information helps characterize and categorize molecules, which is essential for both theoretical comprehension and real-world applications (Beller, 2014). Pharmaceuticals, biology, and materials science are just a few of the domains where carboxylic acids, acyl chlorides, esters, and amides are widely used. Accurate nomenclature enables clear communication and makes it easier to get material from scientific publications, databases, and patents. For researchers to design and develop new medications, explore biological processes, and comprehend the reactivity of molecules that are organic, accurate identification of these substances is essential.

We will examine the systematic guidelines developed by the International Union of Pure and Applied Chemistry (IUPAC) for the nomenclature of carboxylic acids, esters, acyl chlorides, and amides across this document. Readers will get a thorough understanding of how to name and analyze compounds from these important categories of organic compounds by memorizing these rules and examples.

4.2. NOMENCLATURE OF CARBOXYLIC ACIDS

Organic substances called carboxylic acids have the carboxyl functional group (-COOH). To enhance unity and simplicity in communication, the IUPAC created systematic standards for naming carboxylic acids. The essential guidelines are as follows:

1. *Identify the Parent Chain:* The parent chain is the longest continuous carbon chain that includes the carboxyl group. It forms the basis for the name of the carboxylic acid. The suffix "-oic acid" is used to indicate that it is a carboxylic acid.
2. *Numbering the Parent Chain:* The carboxyl carbon (C=O) should be given the lowest number feasible when counting the carbon atoms in the parent chain. commence counting from the very end of the chain if the carboxyl group is there.
3. *Locating Substituents:* If the carboxylic acid has substituents, number the substituents according to where they are located in the parent chain. To distinguish numbers from the names of substitutes, use hyphens. The substitutions should be given the lowest numbers available when numbering.
4. *Naming Substituents:* Alkyl groups are constituents that are joined to the carboxylic acid molecule. For instance, "methyl" refers to a methyl group ($-CH_3$). If there are many substitutes, list them alphabetically while excluding any prefixes like "di-" or "tri-" for ease of use.
5. *Indicating Functional Groups:* Identify the additional functional groups as substituents using the proper prefixes or suffixes in accordance with their functional group nomenclature if they are present in the carboxylic acid molecule.
6. *Common Names:* Some carboxylic acids have widely used common names that are either well-known or have historical importance. For instance, the common names for acetic acid (CH_3COOH) and formic acid ($HCOOH$) are frequently used (Beller, 2014).

Following the IUPAC recommendations, here are some instances of naming carboxylic acids that take into account both saturated and unsaturated structures:

Saturated Carboxylic Acid

Example: CH_3CH_2COOH

In this instance, the chain has two carbons and is finished with a carboxylic acid group. Find the longest carbon chain as the first step. Ethane makes up the parent chain. Determine the functional group of the carboxylic acid as the second step.

The -COOH group is a carboxylic acid. Assign the base name in step three. Ethane is known simply as "ethan." For the carboxylic acid group, add the prefix "-oic acid" (Figure 4.1).

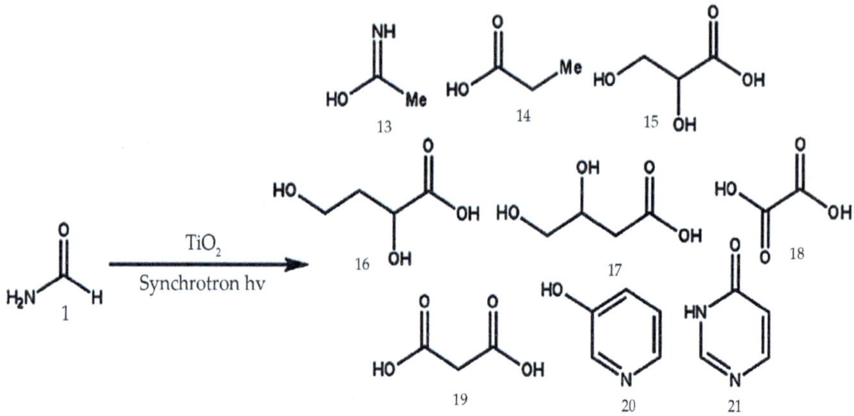

Figure 4.1. Saturated carboxylic acid.

Source: *Saladino Raffaele, Brucato, De Sio Antonio, Botta Giorgia, Pace Emanuele, & Gambicorti Lisa, (2011); Scientific Figure on ResearchGate.*

Count the carbon atoms. The numbering starts at that carbon because the carboxylic acid group is at the end, giving us ethanoic acid. Ethanoic acid, more generally referred to as acetic acid, is the formal name for this substance.

4.2.1. Saturated Carboxylic Acid with Substituents

Example: $CH_3CH(CH_3)\ CH_2COOH$

Here, the third carbon in the chain has a carboxylic acid group connected to it, and the second carbon has a methyl substituent. Find the longest carbon chain as the first step. Butane makes up the parent chain. Determine the functional group of the carboxylic acid. Assign the base name in step three. Butane is known simply as "butan." For the carboxylic acid group, add the prefix "-oic acid."

Count the carbon atoms. The methyl group is located at carbon number two, and the carboxylic acid group is located at carbon number three. The compound is now known as 3-methylbutanoic acid. 3-Methylbutanoic acid is the formal name for this substance (Beller, 2014) (Figure 4.2).

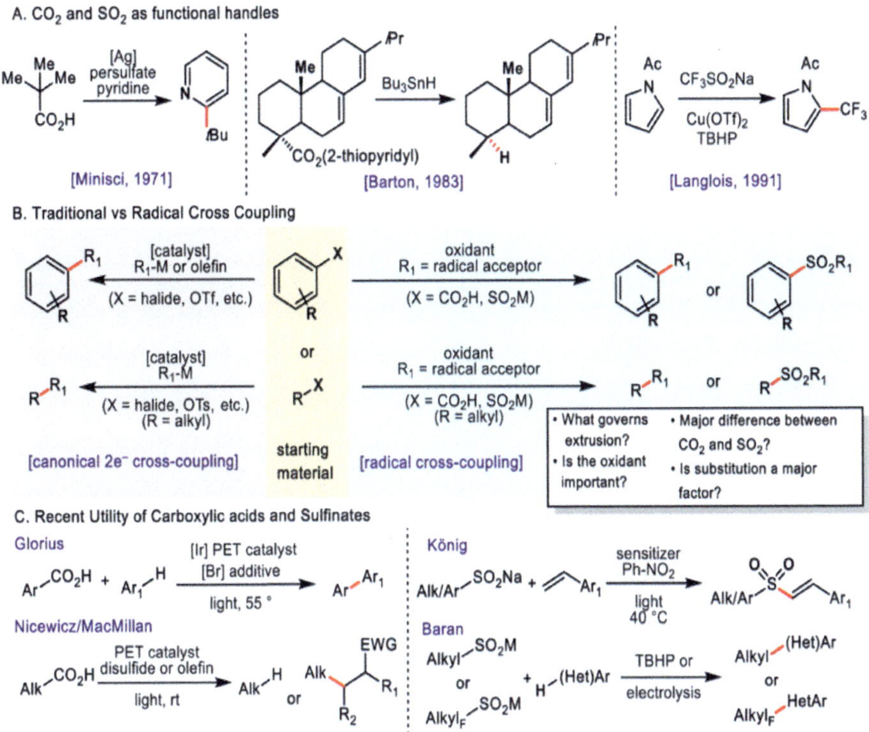

Figure 4.2. Saturated carboxylic acid with substituents.

Source: *Dos Passos Gomes Gabriel, Wimmer Alexander, Spry Joel, König Burkhard, & Alabugin Igor, (2019); Scientific Figure on ResearchGate.*

4.2.2. Unsaturated Carboxylic Acid

Example: $CH_3CH=CHCOOH$

There is a double bond between the second and third carbon atoms in this three-carbon chain, and a carboxylic acid group is added at the end. Find the longest carbon chain as the first step. Propene is in the parent chain. Determine the functional group of the carboxylic acid. Assign the base name in step three. The short form of propene is "propen." For the carboxylic acid group, add the prefix "-oic acid." Count the carbon atoms. The third carbon contains the carboxylic acid group. Propenoic acid is the formal name for this chemical (Nishihara, 2020).

The carboxyl group may not be the most essential functional group for naming purposes if the carboxylic acid is a component of a more complicated structure, it is crucial to mention. In these situations, it may be designated as a substitute by adding the prefix "carboxy-" or "-carboxylate" (Figure 4.3).

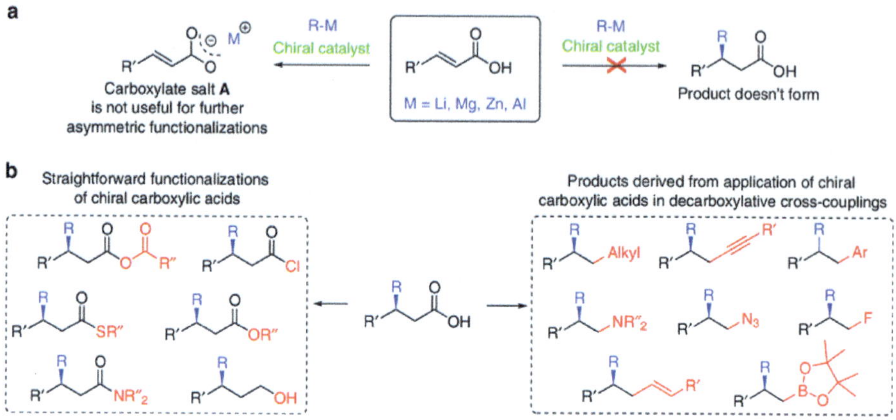

Figure 4.3. Unsaturated carboxylic acid.

Source: *Yan Xingchen & Harutyunyan Syuzanna, (2019); Scientific Figure on ResearchGate.*

Chemists can precisely describe the structure and characteristics of these molecules by adhering to the IUPAC rules for naming carboxylic acids. It is essential for successful scientific communication and the identification of carboxylic acids in various chemical and biological contexts to be able to name them consistently and unambiguously (Beller, 2014).

4.3. NOMENCLATURE OF ESTERS

Esters are organic substances that are produced when alcohol and a carboxylic acid combine. They are generally made up of alkyl or aryl groups for R and R' in their RCOOR' structure. The IUPAC has set criteria for the systematic naming of esters. The IUPAC designations are intended to offer a consistent and clear system of naming chemical substances.

To name an ester using the IUPAC rules, you typically follow these steps:

Define the "acyl" group as the alkyl or aryl group that is linked to the oxygen atom. This is produced from the ester's carboxylic acid portion. To identify the acyl group, change the ending of the appropriate carboxylic acid from "-ic acid" to "-ate." For instance, the acyl group is known as ethanoyl if the carboxylic acid is ethanoic acid.

Identify the "alkoxycarbonyl" group as the alkyl or aryl group joined to the oxygen atom. This is derived from the ester's alcohol portion. After a space, combine the names of the acyl and alkyl carbonyl groups. The name of the acyl group comes first (Nishihara, 2020).

Here are a few instances that show how to name esters with various ester structure types:

4.3.1. Ethyl Ethanoate

The ester ethyl ethanoate is also referred to as ethyl acetate. We can deconstruct the name as follows when using the IUPAC's systematic naming guidelines for esters.The carboxylic acid component, in this case, ethanoic acid, is first identified. The "-ate" suffix is added to the carboxylic acid name in place of the "-ic acid" ending to create the acyl group. As a result, ethanoic acid changes into ethanoyl. The ethanol component of alcohol is the next thing we take into account. We refer to this compound as ethoxycarbonyl because the alkyl group that is connected to the oxygen atom comes from alcohol. The names of the acyl and alkyl groups are then combined, with a gap between them. In this instance, the acyl group is designated as "ethanoyl" and the alkyl group connected to the oxygen atom is designated as "ethoxycarbonyl." Combining them yields the scientific term "ethyl ethanoate." It is important to remember that ethyl ethanoate is frequently referred to as ethyl acetate in both ordinary and scientific contexts. The common names are frequently chosen for simplicity and familiarity, despite the fact that the systematic IUPAC name offers a consistent and accurate means of characterizing the compound.

4.3.2. Methyl Butanoate

An ester with the same name as itself is methyl butanoate. We can deconstruct the name into its constituent parts by using the IUPAC-approved systematic naming procedures for esters.

The carboxylic acid component, in this case, butanoic acid, is first identified. The "-ate" suffix is added to the carboxylic acid name in place of the "-ic acid" ending to create the acyl group. As a result, butanoic acid changes into butanoyl. The methanol component of alcohol is the next thing we take into account. Methoxycarbonyl is the term used since the alkyl group that is connected to the oxygen atom comes from alcohol. The names of the acyl and alkyl groups are then combined, with a gap between them. In this instance, the acyl group is denoted by "butanoyl" and the alkyl group is joined to the oxygen atom by "methoxycarbonyl." These names are combined to form the word "methyl butanoate." The systematic and common names for this ester, "methyl butanoate" and "methyl butanoate," respectively, should be noted. The common name is frequently employed because it is familiar and convenient for communication, whereas the systematic name gives a precise and unambiguous explanation of the compound's structure (Beller, 2014).

4.3.3. PHENYL BENZOATE

The ester phenyl benzoate is frequently referred to by this name. We can deconstruct it in accordance with the naming guidelines for esters in order to comprehend its systematic IUPAC name.

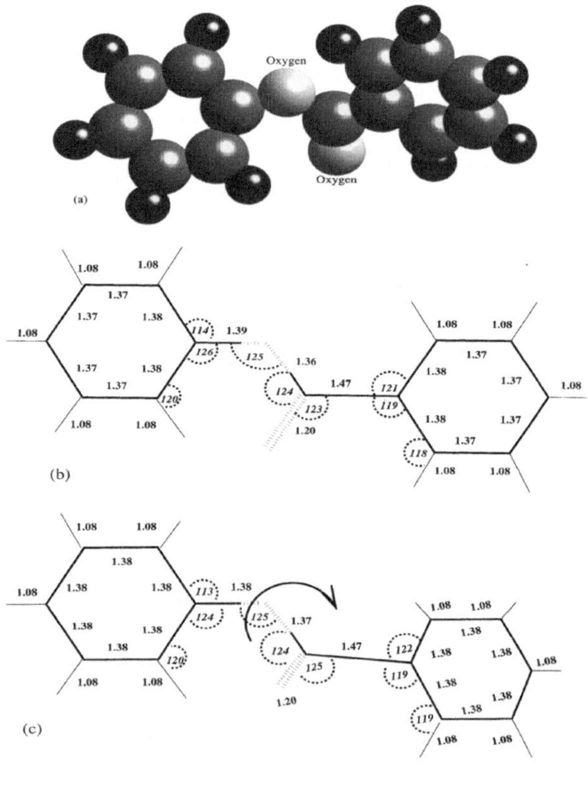

Figure 4.4. Phenyl benzoate structure.

Source: *Adam, Clark, Ackland Gj, & Crain Jason. (1997); Scientific Figure on ResearchGate.*

The carboxylic acid component, in this case, benzoic acid, is first identified. The "-ate" suffix is added to the carboxylic acid name in place of the "-ic acid" ending to create the acyl group. So, benzoic acid changes into benzoyl. We now turn to the phenol-containing alcohol component. In this instance, the term "phenoxy carbonyl" refers to the aryl group that is joined to the oxygen atom. The names of the acyl and aryl groups are then combined; each name being separated by a space. The aryl group connected to the oxygen atom is referred to as "phenoxy carbonyl," whilst the acyl group is known as "benzoyl." The result of combining these names is "phenyl benzoate. It's vital to remember that "phenyl benzoate" is both the systematic and popular name for this ester according to the IUPAC (Nishihara, 2020) (Figure 4.4).

4.4. NOMENCLATURE OF ACYL CHLORIDES

Based on IUPAC recommendations, the systematic naming process for acyl chlorides entails a number of stages to precisely characterize the compound's structure and makeup. Let's explore each step in further depth. Finding the parent carboxylic acid is the first step in determining where the acyl chloride comes from. For instance, carboxylic acid can be used as the basis for identifying the acyl chloride if it is ethanoic acid.

To denote the presence of the chloride functional group (-Cl), the "-ic acid" ending of the carboxylic acid name is replaced with "-yl chloride." This shows that a chlorine atom has been used in place of the parent acid's carbonyl hydrogen (Beller, 2014).

Including substituents: If the carboxylic acid's carbon chain has any substituents, they should be specified as prefixes using the proper numerical locants. The locations of the substituents throughout the carbon chain are specified using numerical locants.

Two examples, both with in-depth explanations: 2-Chlorobenzoyl chloride and 4-Methylbutanoyl chloride (Nishihara, 2020).

4.4.1. 4-Methylbutanoyl Chloride

We first identify the parent carboxylic acid, in this case, butanoic acid, to start the naming procedure. The "-ic acid" ending of the carboxylic acid name must be changed to "-yl chloride" in order to indicate the presence of the chloride functional group. As a result, the core name for this acyl chloride is "butanoyl chloride."

The prefix "4-methyl-" is added to denote the position of the substituent since butanoic acid has a methyl group (CH_3) bonded to the fourth carbon along its carbon chain. The systematic term "4-methylbutanoyl chloride" is created by combining these components and appropriately describes the structure of the acyl chloride with a methyl group on the fourth carbon of the butanoic acid backbone.

4.4.2. 2-Chlorobenzoyl Chloride

The parent carboxylic acid in this instance is benzoic acid. Following the methodical approach, we substitute the "-ic acid" ending with "-yl chloride," giving rise to the core name "benzoyl chloride" for the acyl chloride. Additionally, the second carbon of the benzene ring is joined to a chlorine atom (-Cl). This substitution is denoted by the prefix "2-chloro-." In order to accurately describe the structure of the acyl chloride with a chlorine substituent on the second carbon of the benzene ring in the benzoic acid structure, the systematic term "2-chlorobenzoyl chloride" is obtained.

4.5. NOMENCLATURE OF AMIDES

The IUPAC has established systematic naming guidelines for amides that offer a standardized method for precisely describing the structure and makeup of these substances. Both simple and complicated amide structures must follow these naming guidelines, which include identifying the parent carboxylic acid, switching the "-oic acid" ending for "-amide," and adding the proper prefixes for substituents and functional groups. Let's look at these guidelines in more detail and give examples to show how to properly name amides (Nishihara, 2020) (Figure 4.5).

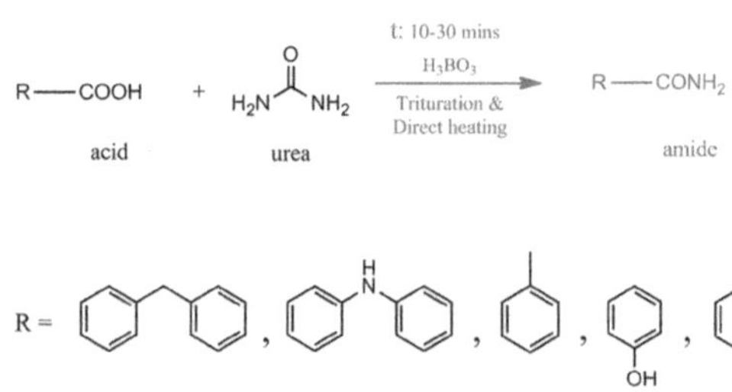

Figure 4.5. Conventional method of amide synthesis.

Source: *Gohil Chiragkumar, & Noolvi Malleshappa, (2021); Scientific Figure on ResearchGate.*

Determine the parent carboxylic acid from which the amide is generated. The first step is to identify the parent carboxylic acid. The amide is named from the parent acid as a starting point.

The "-amide" suffix is added to the carboxylic acid name in place of the "-oic acid" ending to signify the presence of the amide functional group (-$CONH_2$). This modification means that an amine group (NH_2) has taken the place of the parent acid's carbonyl hydrogen (Beller, 2014).

Substituents should be incorporated as prefixes using the proper number of locants if they are attached to the nitrogen or carbon chain of the amide. The positions of the substituents connected to the nitrogen atom or along the carbon chain are indicated by numerical locants.

The parent carboxylic acid in the instance of N-Methylacetamide is acetic acid. The core name "acetamide" is obtained by swapping the "-oic acid" ending for the "-amide," which denotes the amide functional group. The prefix "N-methyl-" designates this substitution because the nitrogen atom has a methyl group (CH_3) connected to it. When these components are combined, the systematic name "N-methylacetamide" appropriately describes the amide with a methyl substituent on the nitrogen atom of acetic acid (Nishihara, 2020).

4.5.1. N, N-Diethylpropanamide

Propanoic acid is the parent carboxylic acid for N, N-Diethylpropanamide. The core name "propanamide" denotes the presence of the amide functional group and is produced by replacing the "-oic acid" ending with "-amide." The prefix "N, N-diethyl-" is used because the nitrogen atom is linked to two ethyl groups (C_2H_5). The amide with two ethyl substituents on the nitrogen atom of propanoic acid is appropriately described by this compound's systematic name, "N, N-diethylpropanamide."

4.5.2. 4-Chlorobenzamide

Benzoic acid is the primary carboxylic acid in the instance of 4-Chlorobenzamide. We get the core word "benzamide" to denote the amide functional group by changing the "-oic acid" ending to "-amide." The prefix "4-chloro-" is used to indicate this substitution because the fourth carbon of the benzene ring has a chlorine atom (-Cl) linked to it. As a result, the systematic name "4-chlorobenzamide" defines the amide in benzoic acid accurately because it has a chlorine substituent on the fourth carbon of the benzene ring.

These illustrations demonstrate the use of methodical amide naming conventions. These guidelines allow us to precisely characterize the structure and makeup of amides, including the presence of functional groups and substituents (Figure 4.6).

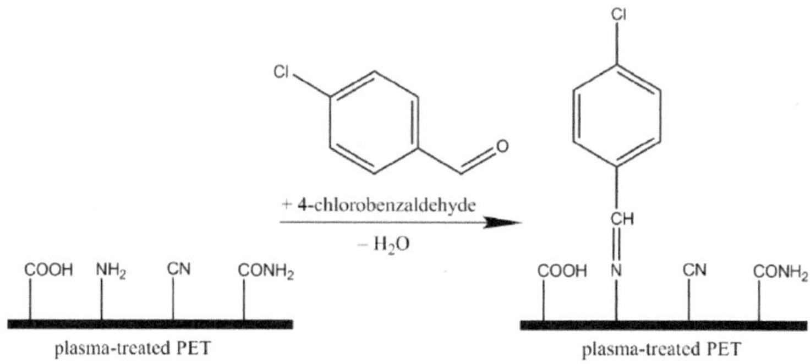

Figure 4.6. 4-Chlorobenzamide.

Source: *Kolar Metod, Mozetic Miran, Stana-Kleinschek Karin, Fröhlich Mirjam, Turk Boris, & Vesel Alenka. (2015). Scientific Figure on ResearchGate.*

4.6. SIMILARITIES AND DIFFERENCES IN THE NOMENCLATURE OF CARBOXYLIC ACIDS, ESTERS, ACYL CHLORIDES, AND AMIDES

To ensure uniform and clear naming of organic compounds, the IUPAC devised a set of systematic criteria for the nomenclature of carboxylic acids, esters, acyl chlorides, and amides. Although the nomenclature of these compounds is somewhat similar, there are also significant distinctions that represent the different functional categories and structural traits of each chemical (Beller, 2014).

The identification of the parent component, which is consistently a carboxylic acid, is one of the key similarities in the name of these compounds. Finding the parent carboxylic acid is the first step in the systematic naming of these compounds. From there, the systematic name of each compound is derived.

All four compound classes address substituents in a similar way. When substituents are present, numerical locants are employed to show where they are connected to functional groups or where they are located along the carbon chain. For instance, substituents are denoting the 'd' by number of locants placed before the relevant functional group or substituent name in carboxylic acids, esters, acyl chlorides, and amides. This makes it easier to pinpoint the substituent's precise location inside the molecule (Nishihara, 2020).

Additionally, all four chemical types can contain alkyl and aryl groups. These groups, which can act as substituents or be a component of the functional group, are named according to the number of carbon atoms in the chain. Alkyl and aryl groups are included because they add diversity to the compounds and enable naming.

Despite these similarities, the name of carboxylic acids, esters, acyl chlorides, and amides differs noticeably and reflects the unique functional groups and properties of each substance. The carboxyl functional group (-COOH) is what gives carboxylic acids their unique properties. In order to properly name a carboxylic acid, the "-e" ending of the parent alkane name must be replaced with the suffix "-oic acid." with instance, the parent component ethane is converted into the derivative ethanoic acid by swapping out the "-e" with the "-oic acid."

In contrast, esters have the ester functional group (-COOR), where R is an alkyl or aryl group. An ester's systematic name is composed of two parts: the name of the alkyl or aryl group, followed by the name of the carboxylic acid, with the suffix "-ate" in lieu of "-ic acid." For example, to create ethyl ethanoate, the prefix "ethyl" is added to signify the alkyl group linked to the oxygen atom and the suffix "-ate" is substituted for "-ic acid" in the original phrase.

The acyl chloride functional group (-COCl) is present in acyl chlorides. By adding "-yl chloride" after the "-ic acid" in the carboxylic acid name, the systematic name of an acyl chloride is created. For instance, replacing the "-ic" suffix on ethanoic acid results in ethanoyl chloride.

The amide functional group (-$CONH_2$) identifies amides. In the systematic nomenclature of an amide, the "-oic acid" ending of the carboxylic acid name is substituted with "-amide." For instance, to denote the presence of the amide functional group, "-amide" is used in place of "-oic acid" to create ethanamide from ethanoic acid.

The prioritizing of functional groups is another distinction. Carboxylic acids emphasize the carboxyl group above other functional groups, whereas esters, acyl chlorides, amides, and acyl chloride-containing compounds prioritize the ester functional group (-COOR), acyl chloride-containing compounds, and amides, respectively.

The location of substituents is the same for all four chemical classes. The particular position of a substituent within a compound is indicated by the number of locants before the relevant functional group or substituent name.

Additionally, each kind of compound has a different nomenclature prefix. Prefixes are typically not used in the names of carboxylic acids. Alkyl or aryl prefixes are used in esters to denote substituents joined to the alkyl or aryl group. Prefixes are often not used in the names of acyl chlorides. In contrast, amides utilize prefixes like "N-" or "N, N-" to denote the presence of alkyl or aryl substituents that have been joined to the nitrogen atom of the amide group (Nishihara, 2020).

In summary, the identification of the parent compound, the use of number locants for substituents, and the presence of alkyl and aryl groups are similarities in the nomenclature of carboxylic acids, esters, acyl chlorides, and amides. Their functional group names, substituent positioning, functional group prioritizing, and nomenclature prefix usage, however, are different. Accurate identification and description of these organic compounds, as well as successful communication and comprehension in the field of organic chemistry, depend on an understanding of these similarities and distinctions (Beller, 2014).

4.7. GUIDELINES TO REMEMBER WHEN NAMING COMPOUNDS FROM THESE CLASSES

It is essential to comprehend and follow a number of fundamental concepts and rules when naming compounds belonging to the classes of carboxylic acids, esters, acyl chlorides, and amides. These rules ensure that the nomenclature is accurate, clear, and consistent, allowing chemists to discuss and identify organic molecules efficiently. Let's examine each idea in more depth:

Parent Compound

In organic chemistry, the classes of carboxylic acids, esters, acyl chlorides, and amides are all descended from carboxylic acids. The parent carboxylic acid, which serves as the basis for the systematic name, is first identified when naming compounds within these classes.

The length of the carbon chain and the presence of the carboxyl group (-COOH) are used to determine the parent carboxylic acid. The number of carbon atoms in the longest continuous chain that has the carboxyl group is used to calculate the length of the carbon chain. The name of the compound is built on this chain. The systematic naming guidelines for each compound class can be used after the parent carboxylic acid has been located.

For carboxylic acids, the systematic name is created by adding the suffix "-oic acid" in place of the "-e" ending of the parent alkane name. For instance, ethanoic acid would be the name of the parent carboxylic acid if it was generated from ethane. When naming carboxylic acids, the carboxyl group comes first. The carbon chain is numbered to show the carboxyl group's position, which is always at carbon 1.

The systematic name for esters is made up of two parts. Alkyl or aryl group name, which denotes the substituent connected to the oxygen atom of the ester functional group, is the first component. The second portion is the name of the carboxylic acid with the suffix "-ate" in place of the prefix "-ic acid." For instance, if the parent carboxylic acid is made from ethane and the alkyl group linked to the oxygen atom is ethyl, the ester would be called ethyl ethanoate.

The systematic term for acyl chlorides is produced by changing the "-ic acid" ending of the carboxylic acid name to "-yl chloride." For instance, ethanoyl chloride would be the name of the acyl chloride if the carboxylic acid was ethanoic acid. When naming acyl chlorides, the acyl chloride group (-COCl) takes precedence.

The "-amide" suffix is added to the carboxylic acid name in place of the "-oic acid" ending to create the systematic name for amides. To show where the nitrogen atom is located, the carbon chain is numbered. Prefixes like "N-" or "N, N-" are frequently used to denote alkyl or aryl substituents joined to the nitrogen atom. For instance, the analogous amide would be if the carboxylic acid were ethanoic acid (Nishihara, 2020).

Functional Groups

In organic chemistry, a substance's functional group is a key factor in influencing the reactivity and other chemical characteristics of the compound. Each class of substance, such as carboxylic acids, esters, acyl chlorides, and amides, has a unique functional group that describes its distinctive properties.

The carboxyl group (-COOH) is what distinguishes carboxylic acids. A carbonyl group (C=O) and a hydroxyl group (OH) are linked to the same carbon atom to form this functional group. The carboxyl group gives carboxylic acids their acidity, which enables them to donate a proton. Carboxylic acids are adaptable substances that take part in a number of chemical processes, including esterification, amidation, and oxidation.

The ester functional group (-COOR), where R denotes an alkyl or aryl group, defines esters. A carbonyl group (C=O) is joined to an oxygen atom, which is then joined to an alkyl or aryl group to form the ester group. Esters are frequently present in essential oils and fragrances and are noted for their pleasant fruity scents. They are created when alcohol and a carboxylic acid combine, removing water in the process. Esters are used in the creation of flavorings, medicines, and solvents among many other things.

The acyl chloride functional group (-COCl) is what distinguishes acyl chlorides, also known as acid chlorides. A carbonyl group (C=O) is joined to a chlorine atom to form the acyl chloride group. Acyl chlorides are extremely reactive substances that are frequently utilized as intermediates in the production of organic compounds. They easily react to form esters and amides with nucleophiles such as alcohols and amines. Acyl chlorides are reactive because of the chlorine atom's strong electronegative charge, which promotes nucleophilic substitution reactions (Beller, 2014).

Amides are identified by the amide functional group (-$CONH_2$), in which a nitrogen atom (N) is joined to a carbonyl group (C=O). Alkyl or aryl groups may be used to further substitute the nitrogen atom. Amides are adaptable substances that are present in many different biological systems and manufactured materials. They are created when an amine or carboxylic acid reacts to generate them. Amides are crucial components of peptides, proteins, and medicinal medicines due to their distinctive structural and electrical characteristics.

Predicting the behavior and reactivity of carboxylic acids, esters, acyl chlorides, and amides requires an understanding of the functional groups that these compounds contain. These molecules have various chemical properties due to the presence of particular functional groups, which enables chemists to customize their synthesis and usage. Researchers can create specialized reactions and functionalize molecules to attain desired features by discovering and using unique functional groups.

Additionally, understanding functional groups enables chemists to forecast how particular compounds will behave in reactions. For instance, the presence of a carboxyl group in carboxylic acids makes them susceptible to reactions such as amidation with amines or esterification with alcohols. The synthesis of esters and amides is made easier by the well-known reactivity of acyl chlorides toward nucleophiles. For chemists to discuss and comprehend the reactivity and behavior of organic molecules, functional groups provide a common language (Nishihara, 2020).

4.7.1. Naming Carboxylic Acids

To name a carboxylic acid, just omit the "-e" ending from the name of the corresponding alkane and add the suffix "-oic acid." For instance, when an alkane like ethane has a carboxyl group, it transforms into ethanoic acid. For the purpose of naming carboxylic acids, the carboxyl group comes first.

4.7.2. Naming Esters

The name of the alkyl or aryl group is the first part of the systematic name of an ester, followed by the name of the carboxylic acid, where the "-ic acid" ending is changed to "-ate." Take the ester produced by the ethyl group and the carboxylic acid known as ethanoic acid, for instance. Ethyl ethanoate is the name of the compound, which combines the ethyl group and the name of the carboxylic acid.

4.7.3. Naming Acyl Chlorides

Acyl chlorides are named by substituting "-yl chloride" for the "-ic acid" ending of the name of the corresponding carboxylic acid. For instance, ethanoyl chloride would be the name of the acyl chloride that corresponds to the carboxylic acid ethanoic acid. It is significant to remember that acyl chlorides can also contain substituted alkyl or aryl groups, which are denoted by numerical locants (Beller, 2014).

4.7.4. Naming Amides

Amides are formed from carboxylic acids, and their names are formed by substituting the "-amide" ending for the "-oic acid" ending. For instance, ethanamide would be the name of the amide that corresponds to the carboxylic acid ethanoic acid. Prefixes like "N-" or "N, N-" before the compound name can be used to denote alkyl or aryl substituents connected to the nitrogen atom in amides.

4.7.5. Substituents

It is important to correctly identify the positions of any substituents present in a chemical when identifying it in organic chemistry. Numerical locants, which stand for the lowest number corresponding to the carbon atom to which the substituent is linked, are used to do this. The use of locants guarantees accuracy and clarity in the systematic naming of compounds and enables chemists to clearly determine the positioning of substituents along the carbon chain.

Alkyl or aryl groups are acceptable substitutes, and they have a big impact on how a product behaves chemically and physically. Aryl groups are formed from aromatic chemicals like benzene, whereas alkyl groups are made up of carbon and hydrogen atoms arranged in a specific chain or branched structure. These groups may be connected to the carbon chain of the parent chemical, act as substituents, or even make up the functional group itself (Nishihara, 2020).

Alkyl and aryl groups are named according to the number of carbon atoms in their chain when they act as substituents. One carbon atom makes up a methyl group, two carbon atoms make up an ethyl group, and three carbon atoms make up a propyl group, for instance. The locant, used to name compounds having alkyl substituents, designates where the substituent is located along the carbon chain. The carbon atom that the substituent is directly linked to must be given the lowest number feasible. This guarantees a methodical and uniform nomenclature.

Aryl groups can act as substituents in compounds in a similar manner. The phenyl group, which is formed from benzene, is the most prevalent example. The locant designates the carbon atom that is immediately joined to the parent chain when names for compounds with aryl substituents are used. To ensure consistency and clarity, it is crucial to give this carbon atom the lowest number feasible.

It is important to remember that the functional group in the molecule might also include alkyl and aryl groups. For instance, in the case of esters, the oxygen atom is joined to the alkyl or aryl group to produce the ester functional group (-COOR). Instead of being treated as a substituent in this situation, the name of the alkyl or aryl group is included as a component of the functional group. The functional group's direct attachment to a carbon atom is indicated by the locant.

Compounds sometimes have both alkyl and aryl groups as additional substituents. In these cases, the positions of each substituent along the carbon chain are indicated the by the number of locants. When the substituents are connected to distinct carbon atoms, the locants are separated by commas. Numerical locants are dropped for subsequent occurrences and the prefixes "di-," "tri-," etc. are used to denote the number of identical substituents if two or more identical substituents are attached to the same carbon atom.

Overall, when naming compounds with substituents, the use of number locants ensures a precise and unmistakable description of the compound's structure. It enables precise research, collaboration, and understanding in organic chemistry by enabling researchers to identify and discuss the locations of substituents along the carbon chain. By adhering to this rule, chemists can create a standardized nomenclature that makes it easier to interpret and anticipate the properties and reactions of compounds (Nishihara, 2020).

Prioritizing the functional group over other substituents is essential for effectively capturing the structure and characteristics of an organic compound when systematically naming it. This rule makes sure that the compound's most important characteristic is reflected in its systematic name. Chemists can convey crucial information about the compound's reactivity, functionality, and chemical behavior by placing focus on the functional group (Beller, 2014).

Since they are the precursors to the families of carboxylic acids, esters, acyl chlorides, and amides, carboxylic acids give priority to the carboxyl group (-COOH). A distinguishing characteristic of carboxylic acids, the carboxyl group is essential to their acidity, hydrogen bonding, and other chemical characteristics. Therefore, the carboxyl group takes precedence in the systematic nomenclature of carboxylic acids. The suffix "-oic acid" is added to the parent alkane name in place of the "-e" ending to indicate the presence of the carboxyl group.

The ester functional group (-COOR) is given priority in the systematic nomenclature of esters. Esters are characterized by the ester functional group, which also gives rise to their distinct characteristics and reactivity. The systematic name appropriately describes the compound's ester nature by placing importance on the ester functional group. The name of the carboxylic acid component is combined with the alkyl or aryl group linked to the oxygen atom to create the ester functional group. Before the carboxylic acid name, the alkyl or aryl group is listed to indicate its location in relation to the ester functional group.

When designating themselves, acyl chlorides give preference to the functional group (-COCl). This compound family is distinguished by the acyl chloride functional group, which also gives unique reactivity and chemical behavior. The systematic name quickly identifies the molecule as an acyl chloride and offers crucial details about its reactivity by giving the acyl chloride functional group priority. The "-yl chloride" suffix is added to the carboxylic acid name in place of the "-ic acid" ending to create the systematic name for acyl chlorides.

The amide functional group (-CONH_2) is given priority in amides' systematic naming. A major factor in the characteristics and reactivity of amides is the amide functional group. The systematic name appropriately identifies the chemical as an amide and provides information about its functional group by prioritizing the amide functional group. When naming amides systematically, the "-oic acid" ending of the carboxylic acid name is changed to "-amide." It is usual to employ prefixes like "N-" or "N, N-" to denote alkyl or aryl substituents joined to the nitrogen atom.

Compound names that prioritize functional groups can more accurately and succinctly convey a compound's structure and attributes. It makes sure that the compound's most important feature is highlighted and gives chemists quick access to information regarding the reactivity and functionality of the substance. This idea allows chemists to communicate and identify substances precisely, promoting successful research projects, teamwork, and knowledge exchange in the area of organic chemistry. The systematic naming of carboxylic acids, esters, acyl chlorides, and amides must prioritize the functional group in order to appropriately describe the compound's structure and convey crucial information about its characteristics and reactivity. Chemists can quickly recognize and comprehend the nature of the chemical by putting the functional group at the beginning of the systematic name, promoting effective communication and furthering organic chemistry research (Nishihara, 2020). Chemists can guarantee consistency and precision in the naming of compounds from the classes of carboxylic acids, esters, acyl chlorides, and amides by adhering to these principles and criteria. This methodical technique enables unambiguous identification of organic substances, effective communication, and a clear description of molecular structures. It is essential to comprehend and put these ideas into practice in order to advance organic chemistry research, instruction, and practical applications across a range of scientific fields.

4.8. IMPORTANCE OF ACCURATE AND CONSISTENT NOMENCLATURE IN ORGANIC CHEMISTRY

For a number of reasons, precise and uniform nomenclature is crucial in organic chemistry. It serves as a common language for scientists to share knowledge, communicate clearly, and comprehend the chemical compositions and features of molecules. Chemists can express precise information on the make-up, structure, and functional groups of organic molecules by following defined nomenclature guidelines. Here, we explore the significance of precise and standardized nomenclature in organic chemistry (Beller, 2014).

First and foremost, precise nomenclature guarantees clarity and prevents ambiguity. Complex architectures with several functional groups and substituents are possible for organic molecules. Chemists can correctly identify and describe these substances with no opportunity for error thanks to a standardized nomenclature scheme. It offers a precise illustration of the molecular structure, enabling scientists to collaborate effectively and researchers to precisely describe their findings.

In addition, nomenclature is crucial for classifying and arranging data about chemicals. Compounds are frequently referred to by their systematic names in databases and scientific literature. Chemical data can be stored, retrieved, and retrieved with efficiency thanks to accurate and consistent nomenclature. It enables researchers to find pertinent information about certain chemicals, advancing organic chemistry research, medication development, and other applications.

Additionally, accurate nomenclature helps explain chemical processes and mechanisms. Understanding the specific structures and functional groups involved is essential for studying reaction mechanisms and forecasting product formation in organic reactions, which involve the conversion of one chemical into another. It is possible for chemists to identify the main transformations and comprehend the underlying chemical principles when beginning materials, intermediates, and products are described using consistent naming conventions.

Correct nomenclature also aids in thinking about safety and legal issues. Organic compounds are subject to strict safety rules and norms in many different industries, including pharmaceuticals, agrochemicals, and food additives. For chemicals to be handled, stored, transported, and used safely, they must be precisely identified and labeled. The appropriate classification, risk assessment, and regulatory compliance of organic chemicals are made possible by proper nomenclature, thereby preserving both the environment and human health (Nishihara, 2020).

Nomenclature also helps with the efficient instruction and learning of organic chemistry. A consistent naming system offers an organized way to comprehend the relationships between molecular structures, functional groups, and compound properties. Aspiring chemists need to understand the basic ideas and principles of organic chemistry. It enables teachers to clearly communicate information, ensuring that pupils gain a firm understanding of the subject. Additionally, using standardized nomenclature gives students the ability to accurately access and analyze scientific literature, which improves their research and analytical abilities.

Correct nomenclature also encourages the repeatability and dependability of scientific experiments. The names of the substances that have been produced or investigated must be specific and unambiguous when researchers publish their findings. This makes sure that the experiments may be repeated and the findings verified by other scientists. The accuracy of chemical identification and characterization is made possible by consistent nomenclature, supporting the objectivity and authority of scientific study (Beller, 2014).

Finally, correct, and consistent nomenclature promotes scientific community communication and worldwide collaboration. Researchers from various nations and cultural backgrounds contribute to the growth of knowledge in the field of chemistry, which is a global science. A standardized nomenclature system offers a universal language that cuts across linguistic and cultural barriers, promoting idea sharing, teamwork in research, and the global dissemination of scientific discoveries.

4.9. IMPORTANCE OF COMPREHENDING THESE COMPOUND CLASSES' NAMING PRACTICES

In organic chemistry, it is crucial to comprehend how carboxylic acids, esters, acyl chlorides, and amides are named. These compound classes serve as key building blocks in the synthesis and study of organic compounds, and appropriate nomenclature facilitates successful fieldwork, clear communication, and accurate representation.

First off, knowledge sharing and good communication between chemists are made possible by understanding the naming conventions. Since organic chemistry is a collaborative field, appropriate understanding and interpretation of the structures and characteristics of carboxylic acids, esters, acyl chlorides, and amides are ensured by uniform naming. It makes it possible to share concepts, research results, and approaches, advancing the field and enabling interdisciplinary cooperation (Nishihara, 2020).

Second, accurate naming makes it easier to identify and categorize organic molecules. A structured framework for describing the molecular structures and functional groups of carboxylic acids, esters, acyl chlorides, and amides is provided by their systematic nomenclature. Chemists can efficiently organize, save, retrieve, and analyze chemical data by adhering to the naming standards, which help them easily distinguish between different molecules. Understanding chemical characteristics, reactivity, and structure-activity correlations are aided by this.

Additionally, comprehension of naming standards helps with predicting and interpreting chemical processes. In synthetic organic chemistry, where it's critical to be able to predict reaction results, carboxylic acids, esters, acyl chlorides, and amides are vital. Assisting in the identification of reaction processes and product prediction, accurate naming sheds light on the functional groups and substituents involved in a reaction. It aids in the creation of new medicines, materials, and other helpful molecules by enabling scientists to build and optimize synthetic routes (Beller, 2014).

Supporting safety considerations in the usage and management of organic substances is proper naming. Understanding the exact compound class and its nomenclature facilitates the assessment of potential risks, toxicity, and environmental effect in sectors including pharmaceuticals, agrochemicals, and fine chemicals. Carboxylic acids, esters, acyl chlorides, and amides must be accurately identified and labeled in order to comply with regulations, assess risks, and ensure that these substances are handled, stored, and transported in a secure manner.

The naming conventions are also crucial for organic chemistry instruction and research. To comprehend the connections between structure and function, students studying the topic need a strong foundation in nomenclature. Learning about organic reactions, processes, and synthesis techniques is made easier by having a firm grasp of the naming standards. Students can acquire pertinent information, navigate scientific literature, and successfully articulate their findings. Accurate naming promotes repeatability in research by offering clear references to particular chemicals and experimental techniques.

Additionally, being aware of the naming conventions encourages the discovery and advancement of novel substances and technology. Accurate naming is crucial for researchers in disciplines like drug development, materials science, and chemical engineering to explain their unique molecules and methods. It makes it easier for scientists to share their discoveries with one another, which helps them expand their bodies of knowledge, conduct more accurate experiments, and verify their findings. By promoting consistency, comparability, and reproducibility in research, precise naming encourages innovation and advancements in these sectors.

The name rules also facilitate successful communication and collaboration among scientists. Researchers in the subject of organic chemistry come from various linguistic and cultural backgrounds. Standardized naming offers a language that can communicate across these divides, enabling scientists to interact internationally and share ideas and research. It facilitates multidisciplinary study, supports knowledge integration, and hastens the advancement of science.

In conclusion, it is critical to grasp the naming conventions for carboxylic acids, esters, acyl chlorides, and amides in organic chemistry for a variety of reasons (Beller, 2014).

CHAPTER

5 Nomenclature of Alcohols, Amines, and Hydrocarbons

Contents

5.1. INTRODUCTION

The area of chemistry known as "organic chemistry" is dedicated to studying carbon-based molecules, which are essential to life and have numerous uses. An important part of organic chemistry is the nomenclature, or system of naming substances.

It offers a structured and methodical way to recognize, group, and explain the characteristics of organic molecules. Nomenclature enables chemists to precisely convey the complexity of organic compounds, facilitating efficient scientific exchange, data interpretation, and teamwork.

It is impossible to overestimate the importance of nomenclature in organic chemistry. Precise and consistent naming conventions ensure clear and straightforward communication among scientists. Chemists can communicate structural information, such as the configuration and connectivity of atoms, functional groups, and stereochemistry, by systematically naming compounds.

Additionally, it sheds light on the physicochemical characteristics of organic molecules. Furthermore, nomenclature promotes effective data management and knowledge exchange across the scientific community by making retrieving information from databases, literature searches, and patent applications easier (McEnally, 2021).

5.2. ALCOHOLS, AMINES, AND HYDROCARBONS OVERVIEW

Three fundamental kinds of organic substances with different structures and properties are alcohols, amines, and hydrocarbons. Alcohols are crucial components of fuels, solvents, and the manufacturing of medicines because they have a functional group called hydroxyl (-OH) connected to a carbon chain.

Amines are substances in which organic groups replace one or more hydrogen atoms in ammonia (NH_3). They are useful in the manufacture of medicines, agrochemicals, and colors. On the other hand, hydrocarbons are the fundamental units of organic chemistry and comprise only carbon and hydrogen atoms. They can be divided into alkanes, alkenes, and alkynes, which have various uses in industrial processes, polymers, and fuels. Alcohols, amines, and hydrocarbons all have distinct nomenclatures that must be understood to study and deal with these significant types of molecules.

5.2.1. Alcohol Structure and Functional Groups

Alcohols are organic substances with a functional hydroxyl (-OH) group linked to a hydrocarbon chain. The position of the hydroxyl group, the length and branching of the carbon chain, and the structure of alcohols can all change. Alcohols are useful molecules used in various applications thanks to the specific chemical and physical features that the hydroxyl group bestows (Boyce, 2018) (Figure 5.1).

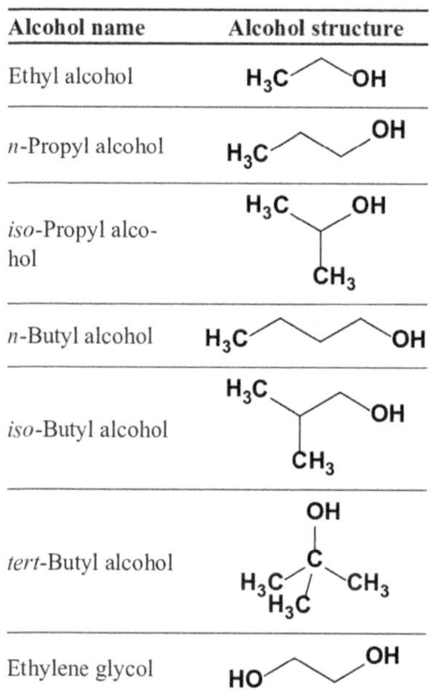

Figure 5.1. Structure of the used alcohols.

Source: *Man Adrian, Gaz, Mare Anca, & Berţa Lavinia, (2017); Scientific Figure on ResearchGate.*

Alcohols' distinctive characteristics, such as hydrogen bonding, solubility, and reactivity, are brought about by the hydroxyl group. Alcohols have higher boiling temperatures and stronger intermolecular interactions than hydrocarbons because hydrogen bonds form between the hydroxyl group's oxygen atom and other molecules' hydrogen atoms. Alcohols are more soluble in polar solvents due to this hydrogen bonding.

Alcohol's functional group is a key factor in their reactivity. It can undergo various chemical processes, such as oxidation, reduction, and substitution reactions. Depending on the circumstances of the reaction, the hydroxyl group can be oxidized to produce aldehydes, ketones, or carboxylic acids. In substitution processes, aldehydes, primary or secondary alcohols, can result from reducing alcohols, although other functional groups, such as halogens, can replace the hydroxyl group. The correct identification, categorization, and naming of alcohols depend on understanding their structure and functional groupings. Alcohols can be named precisely and systematically since the nomenclature rules for these substances consider their structure, substituents, and hydroxyl group position (McEnally, 2021).

5.2.1.1. IUPAC Alcohol Naming

An internationally recognized system for identifying organic compounds, including alcohols, is offered by the International Union of Pure and Applied Chemistry (IUPAC). The systematic identification and arrangement of the carbon chain and the hydroxyl group inside the molecule serve as the foundation for the IUPAC nomenclature for alcohols (Boyce, 2018). To name alcohols using the IUPAC system, several steps are followed. First, the longest continuous carbon chain that includes the hydroxyl group is identified, and it forms the base name of the alcohol. The carbon atoms in the chain are numbered, starting from the end nearest the hydroxyl group.

Next, a suffix is determined based on the number of carbon atoms in the chain. If the chain contains only one carbon atom, the suffix "-ol" is used. For chains with more than one carbon atom, the suffix "-anol" is employed.

The position of the hydroxyl group is indicated by specifying the number of the carbon atom to which it is attached. This number is used as a prefix before the base name.

The IUPAC system ensures a clear and systematic approach to naming alcohols, allowing chemists to communicate these compounds' structure and composition accurately. It facilitates effective data retrieval, classification, and understanding of the diverse properties and reactions of alcohols in various chemical and biological processes.

- *Identifying the Longest Carbon Chain*: When naming alcohols using the IUPAC system, the first step is identifying the longest continuous carbon chain that includes the hydroxyl group. This chain acts as the molecule's backbone and is the foundation for naming the alcohol. Selecting the longest chain to maintain consistency and accuracy in the nomenclature is crucial.

  Identifying the longest carbon chain involves tracing the carbon atoms that form a continuous path from one end of the molecule to the other while including the carbon atom bonded to the hydroxyl group. This chain may be straight or branched, depending on the structure of the alcohol. By selecting the longest chain, we ensure that the base name of the alcohol reflects the main carbon backbone of the molecule.

This step is essential as it sets the foundation for subsequent naming rules, such as identifying substituents and numbering the carbon atoms. Correctly identifying the longest carbon chain provides a systematic and standardized approach to naming alcohols, enabling clear communication and accurate representation of the molecular structure (McEnally, 2021).

- *Determining the Parent Name*: Once the longest carbon chain in the alcohol molecule has been identified, it serves as the basis for determining the parent name. The parent name reflects the number of carbon atoms in the chain and is crucial for accurately describing the size and structure of the compound. Selecting the correct parent name ensures consistency and clarity in nomenclature. It forms the foundation for further naming conventions and provides a systematic approach to denote the main backbone of the alcohol molecule (Boyce, 2018).

  By determining the parent name accurately, chemists can convey important information about the size and complexity of the alcohol, aiding in its identification and understanding within the context of organic chemistry. The parent name sets the stage for the subsequent steps in naming alcohols, allowing for the inclusion of additional substituents and functional groups clearly and logically.

- *Locating and Naming Substituents*: Substituents are additional groups or atoms attached to the carbon chain of an alcohol. Identifying the location of these substituents and naming them correctly is crucial for providing a comprehensive and precise description of the molecule's structure. The position of each substituent is indicated by a number, which corresponds to the carbon atom to which it is attached.

  Locating and naming substituents involves determining the position of the substituent along the carbon chain and assigning it an appropriate name based on the functional group or group of atoms present. The substituents can alter the properties and reactivity of the alcohol, making their accurate identification and naming vital for conveying the complete chemical structure.

By systematically locating and naming substituents, chemists can create an unambiguous representation of the alcohol's structure. This allows for effective communication, facilitates research, and enables the study of structure-activity relationships. Accurate naming of substituents contributes to the overall understanding and classification of organic compounds, promoting consistency and precision in organic chemistry.

- *Assigning the correct suffix:* The suffix of an alcohol name is determined by the number of carbon atoms in the parent chain. The suffix "-ol" is used for chains with only one carbon atom. If the chain contains more than one carbon atom, the suffix "-anol" is employed. Assigning the correct suffix is crucial as it signifies the presence of the hydroxyl group and helps differentiate alcohols from other organic compounds (McEnally, 2021).
- *Examples of IUPAC Naming of Alcohol:* Examples of IUPAC naming of alcohols include methanol (CH_3OH), ethanol (C_2H_5OH), and propanol (C_3H_7OH). In these examples, the number of carbon atoms indicates the parent chain, and the hydroxyl group is named with the appropriate suffix. Any substituents attached to the carbon chain are also located and named using the corresponding numbering system. These examples illustrate the systematic approach employed by the IUPAC system to name alcohols accurately (Boyce, 2018).

5.2.1.2. Common Names of Alcohol

The nomenclature of alcohols also includes common names, which are widely used and have historical significance. Common names are often derived from the source or properties of the alcohol and are not as systematic as IUPAC names. They were widely used before the development of the IUPAC system and are still encountered in various contexts.

- *Historical Perspective on Common Names:* The historical perspective on common names of alcohols offers a glimpse into these compounds' origins and early understanding. Common names often emerged from the substances from which alcohols were first isolated or the properties and used associated with them.

  Other common names were derived from the properties or uses of the alcohol. Methanol, for instance, was initially produced from wood, earning it the common name "wood alcohol." This name highlights its early association with wood-derived sources and its usage as a solvent or fuel.

While common names lack the systematic nature of IUPAC names, they serve as valuable historical markers, providing insights into the development of scientific knowledge and the practical applications of alcohol. These names connect to the early discoveries and observations made by chemists and scientists, reflecting the progression of understanding in the field.

Exploring the historical perspective on common names allows us to appreciate the evolution of chemical nomenclature and the rich history behind the identification and classification of alcohols. While IUPAC names have become standardized, common names continue to carry historical significance and contribute to our understanding of the development and utilization of alcohol throughout history (Boyce, 2018).

- *Examples of Common Names of Alcohols:* Examples of common names for alcohols include methyl alcohol (methanol), isopropyl alcohol (2-propanol), and tert-butyl alcohol (2-methyl-2-propanol). These names often reflect the structural features or origins of the alcohol. For instance, methyl alcohol is derived from the methyl group (-CH_3) attached to the hydroxyl group, while isopropyl alcohol indicates the presence of an isopropyl group (-$CH(CH_3)_2$) in the molecule. These common names have persisted in certain industries, consumer products, and everyday usage, alongside the systematic IUPAC names, due to their familiarity and historical association with specific alcohols (McEnally, 2021).

5.2.2. Nomenclature of Amines

5.2.2.1. Structure and Functional Groups of Amines

Amines are organic compounds that contain a nitrogen atom bonded to one or more carbon atoms. The structure of amines consists of a central nitrogen atom with three substituents attached to it. These substituents can be alkyl groups (R) or hydrogen atoms (H), giving rise to different types of amines, such as primary, secondary, and tertiary amines (Figure 5.2).

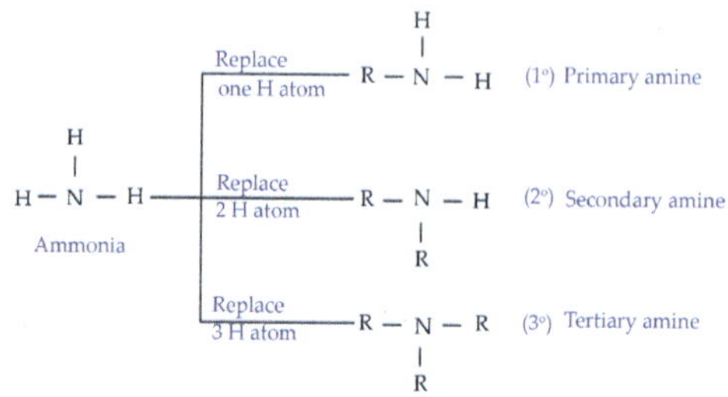

Figure 5.2. Structure of amines.

Source: *Bharadwaj Nithin, (2016); Scientific Figure on ResearchGate.*

The functional group of amines is the amino group (-NH_2), which consists of a nitrogen atom bonded to two hydrogen atoms. This amino group imparts distinct chemical and physical properties to amines. It can act as a base, accepting a proton to form an ammonium ion or undergo various reactions such as nucleophilic substitution, oxidation, and condensation.

The structure and functional groups of amines play a crucial role in determining their reactivity, solubility, and biological activities. By understanding the structure of amines and the functional groups they possess, chemists can predict and manipulate their behavior in chemical reactions and applications.

5.2.2.2. IUPAC Naming of Amines

The IUPAC system provides a standardized method for naming amines based on their structure and substituents. To name amines using the IUPAC system, the substituents attached to the nitrogen atom are identified and named first. The name of the alkyl group is derived from the corresponding alkane by replacing the "-ane" ending with "-yl." The prefix "amino-" is used for primary and secondary amines, while tertiary amines are named N-substituted compounds (Boyce, 2018).

Next, the parent chain is determined based on the longest carbon chain, including the nitrogen atom. The number of carbon atoms in the parent chain is indicated using the appropriate numerical prefix (e.g., "meth-" for one carbon, "eth-" for two carbons). The suffix "-amine" is added to the parent chain name to indicate that it is an amine compound.

In cases where multiple amino groups are present, numerical prefixes such as "di-" or "tri-" indicate the number of amino groups, and the positions of these groups are specified using numbers.

The IUPAC naming system for amines ensures clarity and consistency in identifying and communicating the structures of these important organic compounds. It allows chemists to accurately represent the composition and arrangement of amines, facilitating efficient data retrieval, classification, and understanding in various chemical and biological contexts (McEnally, 2021).

- *Classifying Amines Based on Substituents:* Classifying amines according to their substituents is an essential step in the nomenclature of amines. Primary, secondary, and tertiary amines are the three main categories into which amines can be divided. The number of alkyl or aryl substituents that are joined to the nitrogen atom determines this classification.

  While secondary amines contain two alkyl or aryl groups linked to the nitrogen atom, primary amines only have one. On the other hand, three alkyl or aryl groups are linked to the nitrogen atom in tertiary amines. We can precisely define the structure of amines and distinguish them from one another by categorizing amines according to substituents. Understanding the reactivity and characteristics of amines is also aided by this classification scheme, as the kind and quantity of substituents can affect how they behave in chemical reactions.

- *Naming Primary, Secondary, and Tertiary Amines:* The IUPAC system provides specific rules for naming primary, secondary, and tertiary amines based on their structure and substituents.

  For primary amines with one alkyl or aryl group attached to the nitrogen atom, the alkyl or aryl group is named first using the appropriate prefix (e.g., methyl-, ethyl-, benzyl-). This is followed by the parent chain name, indicating the length of the carbon chain containing the nitrogen atom. The suffix "-amine" is added to the parent chain name to complete the primary amine's IUPAC name (Boyce, 2018).

  Secondary amines contain two alkyl or aryl groups attached to the nitrogen atom. In their IUPAC names, alkyl or aryl groups are named first, followed by the parent chain name indicating the length of the carbon chain containing the nitrogen atom. The suffix "-amine" is added to indicate that it is a secondary amine.

  Tertiary amines have three alkyl or aryl groups attached to the nitrogen atom. In their IUPAC names, all three alkyl or aryl groups are named first, followed by the parent chain name indicating the length of the carbon chain containing the nitrogen atom. The suffix "-amine" is added to indicate that it is a tertiary amine.

These rules ensure a systematic and consistent approach to naming primary, secondary, and tertiary amines, allowing for clear communication and identifying these compounds (McEnally, 2021).

- *Examples of IUPAC Naming of Amines:* Examples of IUPAC naming of amines include ethylamine ($CH_3CH_2NH_2$) as a primary amine, N, N-dimethylamine $(CH_3)_2NH$ as a secondary amine, and N, N, N-trimethylamine $(CH_3)_3N$ as a tertiary amine. In these examples, the substituents attached to the nitrogen atom are named first, followed by the parent chain name indicating the length of the carbon chain containing the nitrogen atom. The systematic use of prefixes and suffixes accurately describes the structure and composition of each amine compound. These examples illustrate the IUPAC naming system's effectiveness in providing unambiguous names for amines, facilitating accurate representation and understanding of these compounds in chemical contexts.

5.2.2.3. Common Names of Amines

In addition to the IUPAC system, amines are often referred to by common names that have historical significance and widespread usage. Common names of amines are derived from various sources, such as the substances they were first isolated from, their characteristic odors, or their applications in specific industries. While the IUPAC system provides a systematic and standardized approach to naming amines, common names continue to be used in certain contexts, including industrial processes, consumer products, and everyday language.

- *Historical Perspective on Common Names:* The use of common names for amines has a rich historical background. Many common names originated from the sources of amines, such as "aniline," derived from the indigo-yielding plant Indigofera anil. Others were influenced by their characteristic odors, such as "putrescine" from decaying flesh or "cadaverine" from cadavers. The historical development of common names reflects the early discovery and identification of amines, often based on their natural sources or distinct properties. While the IUPAC system provides a more systematic and universal approach to naming amines, common names offer insight into these compounds' historical context and early understanding (Boyce, 2018).

- *Examples of Common Names of Amines:* Common names for amines are widely used and recognized in various fields. Examples of common names include "methylamine" (CH_3NH_2), "ethylamine" ($C_2H_5NH_2$), and "pyridine" (C_5H_5N). Methylamine and ethylamine are named based on the alkyl group attached to the amino group. Conversely, Pyridine is a heterocyclic compound containing a six-membered ring with five carbon atoms and one nitrogen atom.

  Common names of amines often reflect their origins or unique characteristics. For instance, "aniline" is a common name for *benzeneamine,* which was initially isolated from indigo plants. "Trimethylamine" refers to $(CH_3)_3N$, a compound with three methyl groups attached to the nitrogen atom.

  Common names of amines have historical and practical significance, as they have been used for a long time and are deeply ingrained in the scientific literature, industrial practices, and common language. While the IUPAC system is the preferred method for systematic naming, familiarity with common names is important for effective communication and understanding in specific domains where they continue to be used (McEnally, 2021).

  It is worth noting that common names of amines can sometimes vary regionally or be subject to different conventions within specific industries or applications. Therefore, it is essential to consider the context and consult the appropriate references when encountering common names of amines.

5.2.3. Nomenclature of Hydrocarbons

5.2.3.1. Structure and Functional Groups of Hydrocarbons

Hydrocarbons are organic compounds composed exclusively of carbon and hydrogen atoms. They are fundamental to 'organic chemistry and are the building blocks for numerous organic molecules. The structure of hydrocarbons is based on carbon atoms forming chains or rings, with hydrogen atoms attached to these carbon atoms (Figure 5.3).

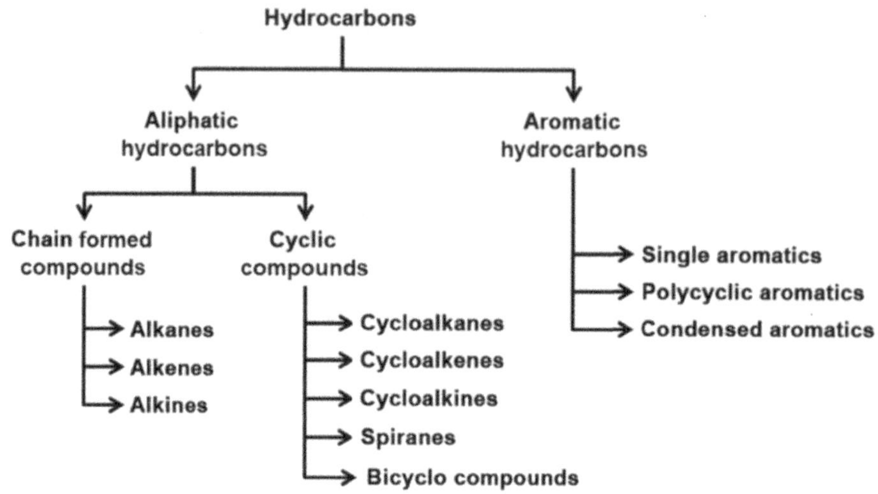

Figure 5.3. Functional groups of hydrocarbons.

Source: *Schmiermund, (2023).*

The presence of multiple bonds or specific arrangements of carbon atoms within the carbon skeleton primarily determines the functional groups in hydrocarbons. For example, alkanes are hydrocarbons with single bonds between carbon atoms and are saturated hydrocarbons. Alkenes contain at least one carbon-carbon double bond, while alkynes possess at least one carbon-carbon triple bond. A benzene ring or other aromatic rings characterize aromatic hydrocarbons (Boyce, 2018).

Hydrocarbons' structure and functional groups determine their physical and chemical properties. The presence of multiple bonds or specific arrangements can influence the reactivity, stability, and behavior of hydrocarbons. Understanding hydrocarbons' structure and functional groups allow chemists to predict and manipulate their properties, developing new compounds with desired characteristics and applications.

5.2.3.2. IUPAC Naming of Hydrocarbons

The IUPAC system provides a standardized method for naming hydrocarbons based on their structure and substituents. The naming of hydrocarbons follows specific rules to describe their composition and arrangement of carbon atoms accurately.

Identifying the Longest Carbon Chain: The first step in naming hydrocarbons is identifying the molecule's longest continuous carbon chain. This is the parent chain and provides the foundation for naming the hydrocarbon. If present, the parent chain is selected to ensure it contains the maximum number of substituents and multiple bonds.

The length of the parent chain is indicated by numerical prefixes such as "meth-" for one carbon, "eth-" for two carbons, "prop-" for three carbons, and so on. The suffix "-ane" is added to the numerical prefix to indicate that it is an alkane (hydrocarbon with single bonds only). For example, a hydrocarbon with a four-carbon parent chain would be named as "butane."

Identifying the longest carbon chain is essential as it determines the base name of the hydrocarbon and sets the foundation for subsequent naming steps. Accurate identification of the parent chain ensures clarity and consistency in the naming of hydrocarbons, allowing for effective communication and understanding among chemists.

By correctly identifying the longest carbon chain, chemists can proceed to determine the substituents, functional groups, and other features of the hydrocarbon, ultimately resulting in a systematic and unambiguous IUPAC name (McEnally, 2021).

Naming Alkanes, Alkenes, and Alkynes: Naming alkanes, alkenes, and alkynes follow specific rules to accurately describe their structures. Alkanes are hydrocarbons with single bonds between carbon atoms, alkenes contain at least one double bond, and alkynes possess at least one triple bond.

When naming these compounds, the number of carbon atoms in the longest continuous chain is identified, and the appropriate suffix (-ane, -ene, or -yne) is added to indicate the type of bond present. Locants are used to specify the position of multiple bonds, substituents, or functional groups along the carbon chain.

By following these naming conventions, chemists can effectively communicate and identify different types of hydrocarbons. The systematic approach allows for easy recognition of the compound's structure, aiding in chemical synthesis, research, and understanding of its properties and reactivity.

Locating and Naming Substituents: Once the longest carbon chain has been identified, the next step in the IUPAC naming of hydrocarbons is to locate and name any substituents attached to the parent chain. Substituents are groups or atoms that replace hydrogen atoms on the parent hydrocarbon chain (Boyce, 2018). The position of substituents is determined by assigning a number to each carbon atom in the parent chain. The numbering starts from the end closest to a substituent or multiple bonds, and the substituent is assigned the lowest possible number. If multiple substituents exist, the numbers are separated by commas and listed in ascending order.

The names of substituents are derived from the corresponding alkyl groups. For example, a substituent consisting of a methyl group ($-CH_3$) is named "methyl," while an ethyl group ($-C_2H_5$) is named "ethyl." Additional prefixes and naming conventions may be applied in more complex cases, such as cyclic substituents or those with branching.

The names of substituents are listed alphabetically before the parent chain name, and their respective position numbers precede them. If multiple substituents exist, numerical prefixes such as "di-" or "tri-" are used.

Accurate locating and naming of substituents ensure precision in the IUPAC naming system for hydrocarbons. It allows chemists to effectively describe hydrocarbon molecules' specific arrangement and composition, enabling better communication and understanding of their structures and properties.

Examples of IUPAC Naming of Hydrocarbons: To illustrate the IUPAC naming system for hydrocarbons, let's consider a few examples. A hydrocarbon with a three-carbon parent chain and a methyl substituent attached to the second carbon atom would be named "2-methylpropane." The methyl group is located on the second carbon atom, and the parent chain is propane (Boyce, 2018). Another example is a hydrocarbon with a six-carbon parent chain, a double bond between the second and third carbon atoms, and a methyl group attached to the fourth carbon atom. The IUPAC name for this compound would be "4-methyl-2-hexene." Here, the substituent is a methyl group located on the fourth carbon atom, and the parent chain is hexene, indicating the presence of a double bond between the second and third carbon atoms. These examples demonstrate how the IUPAC naming system allows chemists to name hydrocarbons based on their structure and substituents precisely. By following the rules and conventions of the system, chemists can generate accurate and unambiguous names for hydrocarbons, facilitating effective communication and understanding within the scientific community (McEnally, 2021).

5.2.3.3. Common Names of Hydrocarbons

Common names of hydrocarbons are alternative names historically used to identify specific hydrocarbon compounds. These names often reflect the source or physical properties of the compounds and are widely recognized in certain contexts, particularly in industrial and commercial applications. While the IUPAC provides a systematic naming system for hydrocarbons, common names continue to be used due to their familiarity and convenience.

Historical Perspective on Common Names: The use of common names for hydrocarbons dates back to early chemical discoveries and the development of the petrochemical industry. Many common names originate from the natural sources of hydrocarbons or their observable properties. For example, "methane" derives its name from the Greek word "methy," meaning "wine," as it was first isolated from distilled alcohol. "Ethylene" comes from "ethy," meaning "sweet," due to its characteristic odor. These historical connections and practical associations contribute to the persistence of common names alongside the more systematic IUPAC nomenclature (Boyce, 2018).

Examples of Common Names of Hydrocarbons: Common names are prevalent for certain hydrocarbon compounds, particularly those with simple structures and wide usage. Here are some examples:

- *Methane (CH_4):* Also known as natural gas, it is commonly used as a fuel for heating and cooking.
- *Ethylene (C_2H_4):* A vital compound in producing plastics and synthetic fibers.
- *Propylene (C_3H_6):* Widely used in manufacturing various plastics, solvents, and chemical intermediates.
- *Butane (C_4H_{10})* is fuel in lighters and portable stoves due to its high energy density.
- *Benzene (C_6H_6):* An aromatic hydrocarbon with wide industrial applications, including producing plastics, detergents, and pharmaceuticals.

Although widely used, these common names can sometimes lead to ambiguity and confusion, highlighting the importance of the standardized IUPAC system for accurate and precise chemical communication.

5.3. COMPARISON AND DIFFERENCES IN NOMENCLATURE

Nomenclature plays a crucial role in accurately identifying and communicating the structures of organic compounds. While alcohols, amines, and hydrocarbons are distinct classes of organic compounds, there are similarities and key differences in their nomenclature.

5.3.1. Similarities in the Nomenclature of Alcohols, Amines, and Hydrocarbons

One notable similarity in the nomenclature of alcohols, amines, and hydrocarbons is using a common base system for naming. The IUPAC system provides a consistent framework for identifying the longest continuous carbon chain in these compounds, which forms the basis for naming. The locations of functional groups or substituents are specified and the carbon atoms in the chain are numbered similarly for all three classes of molecules (Boyce, 2018).

Another similarity is using prefixes and suffixes to indicate the presence of functional groups or unique compound features. For instance, the hydroxyl (-OH) group is denoted by the suffix "-ol" in the case of alcohols. The suffix denotes the amino (-NH_2) group "-amine" similarly for amines. The presence of double or triple bonds determines which suffixes are used to name hydrocarbons, such as "-ane" for alkanes, "-ene" for alkenes, and "-yne" for alkynes (McEnally, 2021).

5.3.2. Significant Nomenclature Differences Among Alcohols, Amines, and Hydrocarbons

Although there are some similarities between naming alcohols, amines, and hydrocarbons, there are also significant differences. Including functional groups in alcohols and amines, which necessitate particular prefixes or suffixes to designate their positions, results in one notable distinction. The name of the alkyl group is followed by a numeral prefix (for example, "hydroxy-") to identify the hydroxyl group in alcohols. When naming amines, the alkyl groups come first, then the amino group, which takes priority over other substituents.

Another important distinction is the inclusion of stereochemical characteristics in the nomenclature of alcohols and amines but not hydrocarbons. The spatial arrangement of atoms in a molecule is referred to as stereochemistry. Prefixes like "R-" and "S-" or "D-" and "L-," respectively, are used to denote the existence of chiral centers or stereoisomers in alcohols and amines. Because they are non-functionalized molecules, hydrocarbons lack stereoisomers and don't need these descriptors.

Additionally, the priority of numbering and naming conventions might vary depending on whether functional groups are present in alcohols or amines. In alcohols, the hydroxyl group is given preference when choosing the carbon chain's lowest possible number. In contrast to hydrocarbons, the amino group takes precedence over other amine substituents, resulting in a distinct numbering scheme.

For appropriate naming and identification of compounds within the groups of alcohols, amines, and hydrocarbons and efficient communication and accurate depiction of their structures in organic chemistry, it is imperative to comprehend these fundamental similarities and distinctions.

5.4. APPLICATIONS AND NOMENCLATURE'S IMPORTANCE OF ALCOHOLS, AMINES, AND HYDROCARBON

5.4.1. Nomenclature's Function in the Documentation and Communication

Nomenclature plays a bigger part in organic chemistry than just giving molecules names. It is a universal language that allows the scientific community to communicate and document effectively. Clarity, precision, and consistency in scientific publications, research papers, patents, and other documentation are ensured by systematically naming compounds using standardized rules, such as those offered by the (IUPAC (Boyce, 2018).

When accurate compound identification and description are crucial in scientific research, communication must be clear and unambiguous. Nomenclature enables scientists to share knowledge about organic molecules' composition and characteristics and properly explain their research findings. By employing standard names, researchers can prevent misunderstanding or misinterpretation of their work, ensuring that their findings are simple to comprehend and easy for others in the field to repeat.

The structure and retrieval of information depend heavily on nomenclature. Compounds are indexed and sorted according to their systematic names in databases, catalogs, and chemical libraries, making it possible to retrieve and cross-reference pertinent information quickly. This makes finding and creating novel compounds easier since it allows researchers to easily look for particular structural or functional groups using the nomenclature (McEnally, 2021).

Additionally, nomenclature helps to classify and categorize substances. Compounds can be organized into groups based on structural similarity and functional groupings by adhering to a systematic naming system.

Understanding organic compounds' characteristics, reactivity, and behavior and investigating structure-activity correlations (SAR) depend heavily on this classification. By comparing molecules with similar names, researchers can spot trends and patterns that enable the development of broad concepts and theories in organic chemistry.

Nomenclature has useful uses in many different industries, in addition to scientific communication and documentation. For example, in the pharmaceutical industry, precise nomenclature is essential for correctly identifying and labeling active components, assuring the safety and effectiveness of medications. Similarly, accurate compound naming in the chemical industry is crucial for quality assurance, product labeling, and legal compliance.

Nomenclature plays a vital role in the communication, documentation, and organization of information in organic chemistry. By following standardized naming conventions, researchers can effectively communicate their findings, classify compounds, and facilitate the retrieval of relevant information. A nomenclature is a fundamental tool for researchers, enabling efficient collaboration, accurate reporting, and advancements in various fields, including pharmaceuticals, materials science, and the chemical industry (Boyce, 2018).

5.4.2. Impact of Accurate Nomenclature in Research and Development

Accurate nomenclature in organic chemistry is significant in research and development, impacting various scientific exploration, innovation, and progress aspects. The following are key ways in which accurate nomenclature influences research and development efforts:

1. *Clarity and Precision:* Accurate nomenclature provides a clear and precise means of identifying and referring to specific organic compounds. This clarity facilitates effective communication among researchers, allowing them to discuss, analyze, and compare compounds confidently. It ensures no ambiguity or confusion when referring to compounds in scientific literature, patents, or research reports.
2. *Structure-Activity Relationship (SAR) Studies:* Nomenclature enables researchers to establish relationships between the chemical structure of a compound and its biological or physical properties. Researchers can organize and analyze data related to

structure-activity relationships by accurately naming compounds. This information is crucial in developing new drugs, materials, and catalysts, as it guides researchers in predicting and optimizing compound properties.

3. *Literature Retrieval and Data Mining:* Accurate nomenclature simplifies the processes of literature retrieval and data mining. Researchers can easily search for specific compounds or classes of compounds based on their systematic names. This capability efficiently retrieves relevant information from databases and scientific literature, enabling researchers to build upon previous work and avoid redundant experiments (McEnally, 2021).
4. *Intellectual Property Protection:* Accurate nomenclature is critical in protecting intellectual property rights. In patent applications, precise and standardized compound names describe the novel compounds or chemical processes being claimed. Accurate nomenclature ensures that the patent application clearly defines the scope of protection, preventing potential infringements and legal disputes.
5. *Reproducibility and Collaboration:* Accurate nomenclature facilitates the reproducibility of research findings. When compounds are named correctly, other researchers can reproduce experiments, verify results, and build upon previous work. It enables effective collaboration among scientists in different laboratories and institutions, as the precise identification of compounds allows for a seamless exchange of information and materials.
6. *Database Integration and Computational Modeling:* Accurate nomenclature enables seamless integration of chemical data into databases and computational modeling tools. Properly named compounds can be accurately processed and analyzed using computer algorithms and modeling techniques. This integration supports high-throughput screening, virtual screening, and molecular modeling studies, accelerating the discovery and optimization of new compounds.

Accurate nomenclature is essential in research and development endeavors. It enhances clarity, enables the analysis of structure-activity relationships, simplifies literature retrieval, supports intellectual property protection, promotes reproducibility, facilitates collaboration, and enables effective integration with computational tools. By employing precise and standardized compound names, researchers can advance scientific knowledge, drive innovation, and make meaningful contributions to various fields of study, including pharmaceuticals, materials science, and environmental research (Boyce, 2018).

5.4.3. Nomenclature's Role in the Chemical and Pharmaceutical Sectors

Nomenclature's critical role in maintaining safety, quality control, regulatory compliance, and efficient communication makes it extremely important in the pharmaceutical and chemical sectors. The following details underline how crucial precise nomenclature is to these industries:

1. ***Labeling and Identification of Drugs:*** Accurate nomenclature is essential in the pharmaceutical sector to identify and label medications. Every active ingredient in a drug must have a defined name, making it possible for patients, pharmacists, and medical professionals to recognize and distinguish between different medications. As a result, medication errors are reduced, and pharmaceutical items are used safely and appropriately.
2. ***Adherence to Regulations:*** Regulating bodies like the Food and Drug Administration (FDA) demand precise nomenclature for the registration and licensing of medicines. Properly named compounds facilitate regulatory compliance by providing a consistent and standardized framework for documenting and reporting drug information. Meeting regulatory requirements is essential for the marketing, distributing, and selling of pharmaceutical products.
3. ***Quality Control and Product Safety:*** Accurate nomenclature is critical in quality control and ensuring product safety in the pharmaceutical and chemical industries. Properly identifying and naming compounds enable precise tracking of raw materials, intermediates, and final products throughout manufacturing. It helps verify substance identity, purity, and potency, minimizing the risk of contamination or adulteration (McEnally, 2021).
4. ***Patent Protection:*** Accurate nomenclature is crucial for protecting intellectual property rights through patents in the pharmaceutical and chemical industries. Precise and standardized compound names are used to define the scope of protection, ensuring that

patented compounds are clearly identified and distinguished from others. Accurate nomenclature helps prevent patent infringements and supports innovation by providing a clear framework for claiming novel compounds or processes.

5. ***Chemical Safety and Handling:*** Accurate nomenclature is essential for ensuring chemicals' safe handling, storage, and transportation. Proper identification and labeling of substances enable workers to understand their properties, hazards, and appropriate handling procedures. It aids in compliance with safety regulations and standards, reducing the risk of accidents, exposure, and environmental harm.
6. ***International Collaboration and Trade:*** Accurate nomenclature in global pharmaceutical and chemical industries enables effective international collaboration and trade. Standardized compound names facilitate communication among scientists, researchers, and industry professionals across different countries and languages. It supports exchanging information, sharing research findings, and global harmonization of regulatory practices.

Accurate nomenclature plays a vital role in the pharmaceutical and chemical industries by ensuring drug identification, regulatory compliance, product safety, patent protection, chemical handling, and international collaboration.

By employing consistent and standardized naming conventions, these industries can uphold quality standards, protect intellectual property, and enhance communication, ultimately benefiting patients, consumers, and the overall advancement of science and technology (Boyce, 2018).

5.5. CHALLENGES AND EXCEPTIONS IN NOMENCLATURES OF ALCOHOLS, AMINES, AND HYDROCARBON

Accurate nomenclature is essential for effective communication and understanding in organic chemistry. However, some challenges and exceptions arise when dealing with complex structures and unique compounds. This section explores some of the key difficulties encountered in nomenclature and the exceptions within the system (McEnally, 2021).

5.5.1. Complex Structures and Naming Conventions

One of the challenges in nomenclature arises when dealing with complex structures. As the size and complexity of organic compounds increase, the naming conventions become more intricate. Long carbon chains, multiple functional groups, and branched structures can make assigning systematic names difficult. In such cases, additional rules and conventions, such as using prefixes or numerical locants, are employed to represent the compound's structure accurately.

Cyclic compounds also present challenges in nomenclature. The naming of rings involves identifying the parent chain, numbering the carbons, and specifying the substituents or functional groups. Multiple rings, fused rings, or bridged structures add further complexity to the naming process. In such cases, using additional prefixes, suffixes, or locants is necessary to denote the arrangement of the rings and their substituents (Boyce, 2018).

5.5.2. Special Cases and Exceptions in Nomenclature

While the IUPAC system provides general rules for nomenclature, certain exceptions and special cases deviate from these rules. One such exception is seen in common names, which are widely used for certain compounds despite not following systematic naming conventions. For example, compounds like acetone and acetic acid have common names that are more commonly used than their systematic IUPAC names.

Another exception is observed in the naming of natural products and derivatives. Natural compounds often possess complex structures and unique functional groups requiring specific nomenclature rules. Common examples include alkaloids, terpenes, and steroids, which have their own naming conventions based on their origins and structures.

Furthermore, some compounds have historical names retained even after introducing systematic nomenclature. For instance, compounds like glycerol and phenol have maintained their historical names due to their long-standing usage and familiarity.

Exceptions also arise in cases where compounds have multiple functional groups or multiple substituents of the same kind. Specific rules are applied to prioritize the numbering and naming of these groups to maintain consistency and clarity in the nomenclature (McEnally, 2021).

Understanding these challenges and exceptions in nomenclature is essential for accurately identifying and communicating complex structures and unique compounds. It highlights the need for continuous refinement and adaptation of nomenclature rules to encompass a wide range of organic compounds and facilitate effective communication in organic chemistry.

In conclusion, in organic chemistry, naming alcohols, amines, and hydrocarbons is crucial. The IUPAC offers systematic naming rules that offer a framework for precisely identifying and describing the structures of organic compounds (Boyce, 2018).

We looked at various nomenclature-related topics, beginning with the fundamental concepts and guidelines for naming alcohols, amines, and hydrocarbons. In the systematic method, the parent chain is identified, functional groups' presence and locations are established, locants are assigned, and prefixes and suffixes denote substituents and functional groups. We also discussed the guidelines for naming them and looked at the various types and classes of chemicals within each group.

It is critical to comprehend and use nomenclature in organic chemistry. It acts as a universal language that makes it possible for the scientific community to communicate, document, and organize knowledge effectively. Clarity, absence of ambiguity, and ease of repeatability are all ensured by accurate and uniform chemical nomenclature.

Nomenclature is used in situations other than academic and research ones. Accurate nomenclature is crucial for drug identification, regulatory compliance, quality control, and intellectual property protection in sectors like pharmaceuticals and chemicals. International cooperation and trade are made easier by proper nomenclature, which enables the handling, storing, and transportation of chemicals in a safe manner (McEnally, 2021).

Both students and experts in the field of organic chemistry must learn the rules and conventions of nomenclature. It improves their capacity to comprehend and communicate structural knowledge, encourages teamwork and creativity, and advances numerous scientific fields.

In organic chemistry, nomenclature is vital for clear communication, classification, and comprehension of organic molecules. Researchers, scientists, and professionals may navigate the complicated world of organic chemistry with accuracy, clarity, and consistency by following defined naming conventions, which will ultimately advance scientific understanding and spur innovation across various study and application fields.

CHAPTER

6 Nomenclature of Nitriles and Carbonyl Groups

Contents

6.1. INTRODUCTION

Nitriles constitute a controversial organic functional group. Although they are generally regarded as harmful, they can be extremely beneficial to a variety of businesses. In this section, we will discuss why these groups are typically toxic, as well as some of their other characteristics. By delving into the structure and synthesis of nitriles, the significance of this functional group will be revealed. A nitrile is a specific functional group used in chemistry. It is a component of organic chemistry due to the presence of carbon in this functional group and its interaction with carbon-containing molecules and compounds (Chang, 2012) (Figure 6.1).

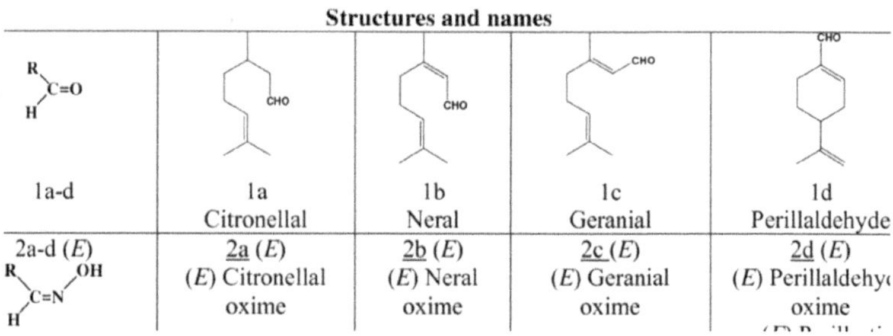

Figure 6.1. Structures of various aldehydes.

Source: *Ouédraogo Igor, Boulvin, Flammang Robert, Gerbaux Pascal, & Bonzi-Coulibaly Yvonne, (2009); Scientific Figure on ResearchGate.*

This term and "cyano-" are frequently used interchangeably in literary works. This prefix is utilized in a variety of contexts to indicate that a molecule contains a nitrile functional group. This prefix indicates that nitrile is present in the molecule. The homologous series nitriles contain a nitrile group.

Cyanocarbons are another term for nitrile-containing chemical compounds. Notably, inorganic compounds containing a nitrile group are referred to as cyanides, not nitriles. The term "cyanides" is derived from the frequently occurring prefix cyano-. The fundamental difference between nitriles and cyanides is that nitriles must be organic, whereas cyanides are commonly used to refer to any chemical compound containing a nitrile group (Wang, 2015).

-CN is the chemical formula for nitrile. However, what does it all mean? A nitrogen atom is triple-bonded to a carbon atom in the nitrile functional group. The carbon will then receive another organic chemistry group, such as a homologous series hydrocarbon. A nitrile is frequently attached to the organic molecule's carbon skeleton.

The nitrile will have a single bond between its carbon and the carbon skeleton of the organic compound and a triple bond between its carbon and nitrogen. How will the carbon and nitrogen atoms in question hybridize? Due to the triple bond and the fact that it will be a linear bond, the two outermost atoms of the functional group will be sp hybridized.

The fact that carbon will have two sp hybridized orbitals is intriguing. As a result, one of the sp hybrid orbitals binds to the organic complex's carbon backbone, while the other binds to its nitrogen. Each of these instances involves the formation of a sigma bond.

For the nitrogen to form a sigma bond with the carbon, it must also undergo sp hybridization. Where do the other two bonds in a triple bond originate from when the sigma bond only yields one? Carbon and nitrogen both have two 2p orbitals. When these two come together, two pi bonds are created. These two additional orbitals finish the triple bond between nitrogen and carbon in nitrile.

Nitriles are pervasive because they are an indispensable element of organic chemistry. Here, we will examine some frequent examples and the contexts in which they occur.

Overall, there are two categories of nitriles. Most individuals associate nitriles with negative connotations. Typically, cyanide and its lethal properties are involved (Wang, 2015).

Cyanide has played a significant role in history for many years. This is a result of the ongoing demand for its products for a variety of applications. In this manner, its significance in culture and the media has increased over time. This spans the Medieval and Modern eras.

On the other hand, there are substances to consider that are advantageous. As a functional organic group, it can be found in a variety of organic chemistry situations (Chang, 2012).

Examples of polymers with significant applications include rubber and adhesive. Nitriles are present in both superglue and rubbers such as nitrile elastomer. Due to its resistance to lubricants and fuels, nitrile rubber is vital in a variety of industries. This is primarily attributable to the nitrile's polymers, which are resistant to a variety of organic hydrocarbon compounds, including fuels. In situations where latex is not used, nitrile-containing polymers may contain another polymer known as nitrile. This is commonly used as a substitute for latex, which can be life-saving for individuals with latex allergies. Consequently, it is commonly found in latex-free laboratories, medical garments, and hospital settings (Figure 6.2).

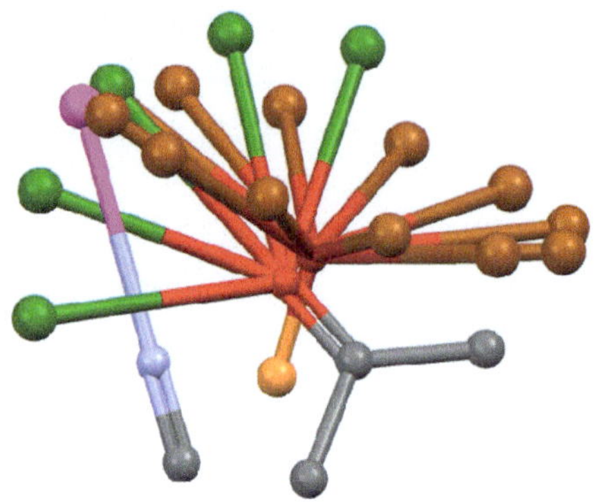

Figure 6.2. Example overlay of nitrile, carbonyl, and phosphate acceptors.

Source: *Taylor Robin, Cole Jason, Cosgrove David, Gardiner Eleanor, Gillet Val, & Korb Oliver, (2012); Scientific Figure on ResearchGate.*

To obtain the IUPAC names for carboxylic acids, the final e of the corresponding hydrocarbon is removed, and -oic acid is appended. Remember that the carbon atom of carboxyl is always C-1. The majority of simple carboxylic acids have more prevalent common names than their IUPAC names. The suffix carboxylic acid is used to identify cyclic carboxylic acids. The suffix -dioic acid or -dicarboxylic acid identifies substances with two carbonyl groups, depending on the context. The fundamental designation of the linked hydrocarbon retains its final (Wang, 2015).

The name of a simple nitrile is formed by prefixing the hydrocarbon's name with "-nitrile." Keep in mind that C-1 represents the carbon in nitrile. Cyclic nitriles are named by substituting -onitrile for -ic or -o acid, as they are commonly believed to be formed from carboxylic acids.

Ketone and aldehyde both contain a carbonyl group. It is believed that aldehydes are the most significant functional group. The group is referred to as the formyl or methanoyl group. Alcohols undergo dehydration to form aldehydes. The carbonyl group is linked to at least one hydrogen atom in aldehydes. The carbonyl group is connected to two carbon atoms in ketones.

Aldehydes and ketones are organic compounds with the carbonyl functional group or C=O. This group's carbon atom has two vacant bonds that can be supplied with hydrogen, alkyl, or aryl substituents. The compound is an aldehyde if at least one of these substituents is hydrogen. If none are hydrogen, the substance is a ketone.

As is typical for compounds with two or more functional groups, one functional group is highlighted while the other is classified as a substituent. In compounds containing both functional groups, ketones are referred to as "oxo" substituents because aldehydes have a higher priority than ketones. The aldehyde functional group does not require a location number, whereas the ketone typically does.

Due to the likelihood that the aldehyde functional groups in dialdehydes occupy the extremities of the parent chain, the placement numbers for both carbonyls are eliminated. At the end of the primary chain name, an additional -dial is appended (Chang, 2012).

When an alcohol functional group is present alongside an aldehyde or ketone in a molecule, the IUPAC system prioritizes the carbonyl nomenclature. The carbonyl has been assigned the lowest conceivable location number, and the proper nomenclature affix is present. Regarding alcohols, OH is referred to as a hydroxyl substituent. Nevertheless, the l from hydroxyl is typically omitted.

When an aldehyde or ketone is present in a molecule with an alkene functional group, the carbonyl nomenclature is prioritized by the IUPAC system. Carbonyl is given the exact nomenclature prefix and the lowest location number. When carbonyls are added to an alkene, the following results:

The formula below establishes the existence of an alkene. (Omit the -ane for the longest carbon chain) (The suffix "-en" implies that an alkene is present.) The carbonyl location number is (either -one or -anal) if a ketone is present (Wang, 2015).

Keep in mind that the carbonyl has precedence, therefore its location number should be as low as possible. If necessary, the alkene's cis/trans or E/Z nomenclature must be specified.

The carbonyl group is attached to a carbon and a hydrogen atom in aldehydes, but two carbon atoms in ketones. In carboxylic acids, a carbonyl group is connected to the carbon, hydrogen, or oxygen of the hydroxyl group (-OH), whereas in amides and acyl halides, it is related to the carbon, hydrogen, or nitrogen of the $-NH_2$ group or to halogens. Carboxylic acids react to generate esters and anhydrides (Chang, 2012). Plants and animals both contain aldehydes, ketones, and carboxylic acids. They are required for all metabolic activities in living organisms. They improve nature's flavor and scent. The scents of salicylaldehyde, cinnamaldehyde, and vanillin, generated from vanilla seeds, are all pleasant.

They are used to improve the flavor of a variety of medical and culinary goods. Certain members of these families are used as solvents in the manufacture of adhesives, pigments, polymers, perfumes, plastics, textiles, and other products (for example, acetone).

Aldehydes and ketones are more generally known by their common names than by their IUPAC nomenclature. The bulk of aldehyde common names are created from carboxylic acid common names by replacing the -ic suffix with an aldehyde. Furthermore, the names are translations of the Latin or Greek nomenclature for the source of the acid or aldehyde. The location of the substituent along the carbon chain is denoted by Greek letters such as a, b, g, and d. The first carbon instantly enters the aldehyde group, followed by the second, and so on.

Traditionally, a ketone is described as two alkyl or aryl groups connected to a carbonyl group. Greek letters beginning with aa represent the location of carbon atom substituents near the carbonyl group. Some ketone molecules have historically common names; the most fundamental dimethyl ketone is acetone. To distinguish between alkyl phenyl ketones, the name of the acyl group is usually appended to the word "phenone."

By swapping -e for -al or -one, the names of open-chain aliphatic aldehydes and ketones are derived from the names of the corresponding alkanes. Ketones are numbered from the carbonyl-proximal end, whereas aldehydes are numbered from the carbon of the aldehyde group. Numbers matching the alphabetical order of the substituents are used to denote the substitution sites in the carbon chain. The same is true for cyclic ketones, where the carbonyl carbon is represented by the number 1. When an aldehyde group is linked to a ring, carbaldehyde is added to the cycloalkane's legitimate name. The carbon atom attached to the aldehyde group is the ring's initial carbon atom. The most fundamental aromatic aldehyde is benzenecarbaldehyde, which has an aldehyde molecule linked to a benzene ring. IUPAC, on the other hand, adopts the common name benzaldehyde. Substituted benzaldehydes are among the extra-aromatic aldehydes (Chang, 2012).

Three sigma (s) bonds are created as a result of the carbonyl atom's sp^2 hybridization. The fourth valence electron of carbon remains in its p-orbital and establishes a p-bond with oxygen by interfering with the p-orbital of an oxygen atom. In addition, the oxygen atom has two unpaired electrons. As a result, the carbonyl carbon and its three associated components can be found both above and below the p-electron cloud. A trigonal coplanar structure is projected to have a bond angle of about 120°.

The double bond between the atoms is polarized because oxygen has a stronger electronegativity than carbon. As a result, the Lewis base's nucleophilic core (carbonyl oxygen) and Lewis acid's electrophilic core (carbonyl carbon) are switched. Carbonyl compounds have higher dipole moments and are more polar than ethers. The carbonyl group's high polarity is explained by the resonance between neutral (A) and dipolar (B) structures (Wang, 2015).

At room temperature, the gas methane exists. Ethanol is a very flammable liquid. At room temperature, other aldehydes and ketones are either liquids or solids. Aldehydes and ketones have greater boiling points than similar molecular weight hydrocarbons and ethers. The molecular bonds between aldehydes and ketones are weak due to dipole-dipole interactions. Their boiling temperatures are lower than those of alcohols of the same molecular weight due to the absence of intermolecular hydrogen bonding. The compounds having molecular weights between 58 and 60 are given in decreasing order, along with their boiling temperatures.

Because propanone, ethane, and methane form hydrogen bonds with water, they are miscible at any concentration. In contrast, the solubility of aldehydes and ketones drops considerably as the alkyl chain lengthens. Aldehydes and ketones are all relatively soluble in organic solvents such as benzene, ether, methanol, and chloroform. The odor of the lowest aldehydes is strong and unpleasant. The smell grows stronger and more fragrant as the size of the molecules rises. Flavors and fragrances are created using a range of naturally occurring aldehydes and ketones (Chang, 2012).

Organic compounds having a nitrile functional group are known as nitriles. These functional groups, which consist of a triple-bonded carbon atom and a nitrogen atom, can generate polar compounds with high dielectric constants. Although the functional group is more common in organic molecules, it may also be found in inorganic compounds like cyanides. The structural properties of isonitriles, cyanates, and thiocyanates all impact their functioning. Nitriles are useful molecules with several chemical applications. In organic chemistry, nitriles are often utilized as solvents and synthons. Acetonitrile, the smallest organic nitrile, is often employed as an organic solvent or in reverse-phase chromatography. As a result of improvements in polymer chemistry, several major nitrile-containing polymers, such as polyacrylonitrile, poly(acrylonitrile-co-butadiene-co-styrene), poly(styrene-co-acrylonitrile), and nitrile rubber, have been produced. These polymers may be modified to perform new activities, resulting in the creation of novel macromolecules with fascinating features. In laboratory gloves, nitrile rubber is preferred over latex. The pharmaceutical business is the largest nitrile market (Wang, 2015). The nitrile functional group is found in a variety of drugs, some of which are still in clinical trials. Carbonyl bioisostere, hydroxyl and carboxyl surrogates, azomethine-water isostere (for cyanoquinolines and cyano pyridines), carboxyl transition state analog, halogen bioisostere, and ADME-toxicology profile enhancer are some of the compounds in this class. Because the nitrile moiety in medicines performs various biological roles, these chemicals are used to treat a wide range of medical disorders. A recent study looks at over thirty FDA-approved nitrile-containing medications, as well as the advantages of the functional group's improved pharmacokinetic profile, lower drug resistance, and greater binding affinity. Anastrozole, an antiarrhythmic drug used to treat angina, and tofacitinib, a Janus kinase inhibitor used to treat rheumatoid arthritis, psoriatic arthritis, and ulcerative colitis, are two examples.

As stated previously, the nitrile group functions as a bioisostere for several functional groups, including halogens and carbonyls. By blocking metabolically unstable sites with nitrile activity, drug molecules can become more metabolically stable. The proposed chemical significantly affects the bioactivity and toxicity of nitriles. Toxic nitriles can induce neurological, hepatic, cardiovascular, renal, and gastrointestinal disorders. Cyanogenic substances that can generate hydrogen cyanide (HCN) are exceedingly hazardous. Numerous natural nitriles have been identified as cyanogenic glycosides in more than 2500 plant species, including significant food sources. These compounds have the potential to emit HCN after ingestion and metabolism, making them potentially hazardous (Chang, 2012).

Since the discovery of the first cases, interest in natural nitriles has increased significantly, and there are now numerous evaluations of such nitriles in the scientific literature. There may be nitrile-containing compounds in plants, insects, fungi, bacteria, corals, and other aquatic organisms. Based on 60 references, a 1981 study reported the discovery of compounds with cyano or isocyano functionality in plants, animals, and fungi. A second publication from 1990 compiled the then-current knowledge on nitriles in natural products. This study described the formation of nitriles, as well as their decomposition and detoxification, and listed all known chemicals. In a 1999 book, all nitrile natural products discovered between 1990 and 1999 were categorized according to the degree of carbon hybridization that gave origin to the nitrile, continuing the work of previous studies. At the time, over 120 natural substances, excluding cyanogenic glycosides, contained nitriles. In actuality, a 2021 review of cyanogenic glycosides included the identification of 112 cyanogenic glycosides to date. To continue previous endeavors, this study will characterize natural compounds discovered between 1998 and 2021 that contain at least one nitrile functional group. Due to the absence of compounds containing the isomeric isonitrile functional group, the reader is referred to a recent comprehensive review. This article does not discuss labile natural substances such as glucosinolates, which can transform into nitriles. They are the subject of yet another evaluation (Wang, 2015).

6.2. CARBONYL COMPOUNDS NOMENCLATURE

6.2.1. Ketones (Alkanones) and Aldehydes (Alkanals)

Ketone R groups (R1 and R2) must each contain a C atom directly connected to the carbonyl group; neither R group may contain an H atom. The root identification of a compound is determined by the length of the longest and most continuous carbon chain containing the C=O group. Since ketones are formed by oxidizing the relevant 2° alcohol, they are given the same scientific nomenclature.

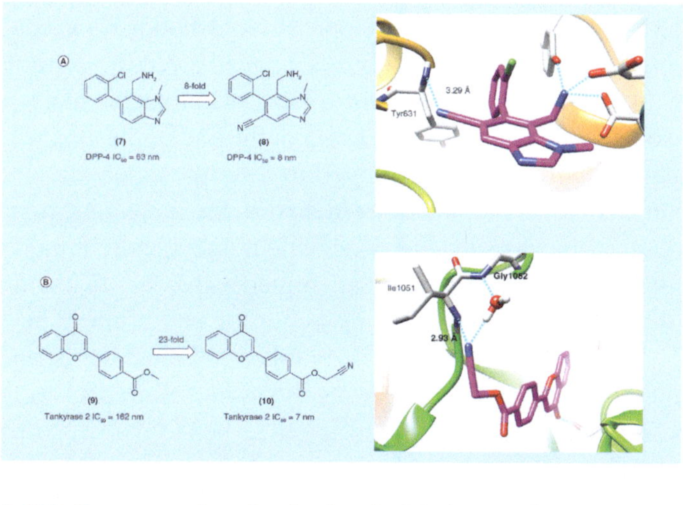

Figure 6.3. Nitrile groups forming hydrophobic interactions.

Source: *Yuanxun Wang, Du Yunfei, & Huang Niu, (2018); Scientific Figure on ResearchGate.*

The suffix "one" replaces the prefix "ol" in the name of the previous alcohol to produce the word "ketone." Similar to how we determine the position of the C-OH in the precursor 2° alcohol, we append a number to the root "alkanone" name to indicate the position of the C=O group in the chain (Figure 6.3).

Like aldehydes and carboxylic acids, ketones lack common names. In contrast to their scientific designations, simple ketones are known by a multitude of common names. As seen in the image above, the prefix aceto in acetophenone is derived from the prefix acet in acetic acid. The prefix acet is also used to distinguish acetone, a member of the well-known ketone family (Chang, 2012).

To differentiate ketones, the IUPAC nomenclature uses the suffix -one. The position of the carbonyl function of a ketone within a chain or ring is typically denoted by a location number. Utilizing standard chain numbering, commencing at the end nearest to the carbonyl group, the chain is numbered. Because there is only one potential location for a ketone carbonyl function, simple ketones such as propanone and phenylethanone do not require locator numbers. The common names of ketones are derived from the two alkyl groups that are attached to the carbonyl group, followed by the suffix -ketone. The names of related alkyl groups are displayed in alphabetical order (Wang, 2015).

The general structure of an aldehyde is similar to that of a ketone. In contrast, aldehydes require at least one H-C=O bond. As a consequence, the C=O group of aldehydes is always located near the terminal carbon atom of an alkane molecule. In contrast, the C=O molecule of a ketone must always be located in the middle and never at the end of the alkane chain. The term "alkanals" is frequently employed to describe aldehydes. Because 1° alcohols can be oxidized to produce aldehydes, the systematic nomenclature for 1° alcohols is the basis for the systematic nomenclature for aldehydes. In the name of the aldehyde product, "al" replaces the "ol" that concludes the precursor 1° alcohol. Because it is always at the end of the chain, the C=O group does not require a number. The number "1" must be appended to the name of the 1° alcohol because all alcohols are referred to as "alkanols" regardless of whether the OH group is on carbons C1 or C2 or any other carbon in the chain.

Why are aldehydes and ketones recognized as distinct classes but 1° and 2° alcohols are not? This is likely due to the discovery by early scientists that aldehydes can tolerate many essential chemical reactions while ketones cannot. Aldehydes are converted into carboxylic acids with the same chain length as previously indicated. When ketones are oxidized, the R group typically dissipates and the C=O chain is shortened.

Although aldehydes can be oxidized, the chain length is unaffected because the C=O group is located at the end. However, since the preponderance of additional aldehyde and ketone chemical reactions are identical, they would not have been separated into two categories if they had been discovered more recently. If the ketone nomenclature system were utilized, aldehydes would be referred to as 1-alkanones; however, this concept has not yet gained widespread acceptance.

The common terms for simple aldehydes and the carboxylic acids to which they oxidize are identical. As with carboxylic acids, Greek letters designate the positions of substituents on aldehydes (Chang, 2012).

The IUPAC nomenclature system allocates aldehydes to the suffix -al. For instance, $H_2C=O$ contains methanal, also known as formaldehyde. It is unnecessary to include an aldehyde carbonyl group in the designation, as it must always be located at the end of a carbon chain and is always designated first. IUPAC maintains the common names for several simple carbonyl-containing compounds (Figure 6.4).

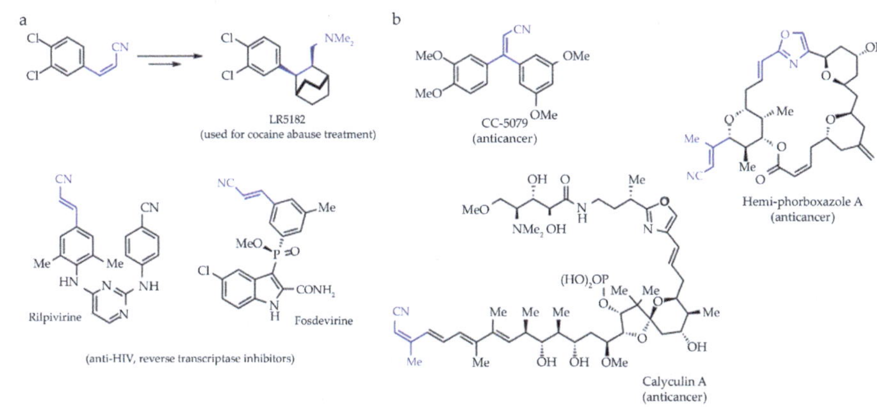

Figure 6.4. Biologically active compounds.

Source: *Mu Yucheng, Nguyen Thach, Koh Ming Joo, Schrock Richard, & Hoveyda Amir, (2019); Scientific Figure on ResearchGate.*

Aldehydes and ketones can also be named using a standard method. Similar to carboxylic acids, aldehydes contain the suffix -aldehyde after their parent chain names. Greek letters are typically used to designate carbon atoms adjacent to the carbonyl group in aldehyde common names. Alpha is the atom closest to the carbonyl function, followed by beta, etc.

If an aldehyde moiety (-CHO) is added to a ring's name, the resulting suffix is -carbaldehyde. This portion's carbon will be allocated to position number 1 on the ring (Wang, 2015). The IUPAC nomenclature system allocates the prefixes al for aldehydes and one for ketones to these classes. For instance, $H_2C{=}O$ contains methanal, also known as formaldehyde. Because an aldehyde carbonyl group is always located at the end of a carbon chain, it automatically occupies position 1 and determines the direction of the numbering. A locator number identifies a ketone carbonyl function that can exist anywhere in a chain or ring. Standard chain numbering begins at the end closest to the carbonyl group. The carbonyl group in cyclic ketones is assigned position 1, and this number is not indicated in the name unless there are numerous carbonyl groups. Immediately consult the IUPAC nomenclature guidelines if you have any questions (Chang, 2012).

In the figure below, examples of IUPAC names are indicated in blue. The derived names are rendered in black, while the common names are rendered in red. Greek letters are frequently employed in common names to represent carbon atoms close to the carbonyl group. Alpha is the atom closest to the function, followed by beta, etc. Two close-knit groups of atoms make up ketones; one is designated etc. and the other," etc.

Propanone and phenylethanone, the first two examples in the right column, are two extremely basic ketones that do not require a locator number because there is only one conceivable location for a ketone carbonyl function. Because aldehyde functions must occupy the ends of carbon chains, locator numbers are not provided, as they are for dialdehyde. The examples of hydroxy butanal, propenal, and oxopropanal (second and third from the top, left column) illustrate which IUPAC prefixes have name precedence. In every instance, the aldehyde function serves as the nomenclature suffix and is ranked higher than alcohol, alkene, and ketone. All remaining functional categories are replacements. Because ketones have a lower suffix priority than aldehydes, the "oxo" substituent terminology is rarely employed. Common are simple substitutions with a carbonyl group. These categories are collectively known as acyl (Wang, 2015).

Like aldehydes, ketones are named according to IUPAC rules, with "one" replacing "e" in the name of the hydrocarbon chain with the longest chain length. In acyclic ketones, the carbon with the lowest locant value is the ketonic carbon. As shown below, pentan-2-one is used instead of pentan-

4-one to refer to a straightforward five-carbon ketone. According to IUPAC standards, the locant value may also appear before the parent's name to generate a new designation, such as 2-pentanone.

In cyclic ketones, the carbonyl carbon is numbered first. As the numbering descends, the value of the succeeding substitute on the ring decreases. For instance, "3-methyl cyclopentanone" rather than "4-methyl cyclopentanone" is used to designate a cyclic ketone with five carbons and a methyl group.

If the primary chain contains functional groups with a higher priority, such as acids, esters, and aldehydes, the ketone group is considered a substituent. In these instances, the locant numerals are preceded by "oxo," which denotes the ketone group. Examples of 3-oxo butanoic acid and 3-oxopentanal are provided.

If the progenitor chain does not contain a ketone substituent, the group is classed as either acyl or alkanoyl. The main ring of 4-acetyl benzaldehyde lacks a ketone group, as indicated in the picture below. As a result, the term acetyl is formed by suffixing benzaldehyde with acetyl.

The carbon atom in the carbonyl group displays sp^2 hybridization and a trigonal planar shape in both aldehydes and ketones. Two sp^2 orbitals on the carbonyl group's carbon atom are used to build interactions with other carbon or hydrogen atoms inside molecules. The last sp^2 hybrid orbital is covalently connected to the oxygen atom. To create the double bond, the unhybridized p orbital on the carbon atom in the carbonyl group crosses over a p orbital on the oxygen atom.

Because oxygen is far more electronegative than carbon, the carbonyl group's C=OC=O bond is polar, comparable to the C=OC=O bond in carbon dioxide. A Lewis base interacts with the carbon atom at the positive end of the polar C=OC=O bond to generate an unstable intermediate, which then undergoes one or more structural modifications to yield the final product in a sequence of reactions involving aldehydes and ketones (Chang, 2012).

The production of aldehydes and ketones demonstrates the relevance of molecular structure in organic compound reactivity. By oxidizing an alcohol, a carbonyl group can be formed. A carbon atom is oxidized when a carbon-hydrogen link in an organic molecule turns into a carbon-oxygen bond. This carbon atom is diminished when a carbon-hydrogen bond substitutes a carbon-oxygen bond. Remember that unless it is an element or associated with fluorine, the oxidation value of oxygen is -2. When hydrogen is unbonded from metal, its oxidation number is usually 1. Because the oxidation number of carbon is not constant, it must be determined algebraically using the bound elements and the total charge of the molecule or ion. A carbon atom's oxidation number is higher when it is connected to an oxygen atom and lower when it is linked to a hydrogen atom. This should be consistent with your evaluation of the polarity of the C-O and C-H bonds (Wang, 2015).

Aldehydes are formed when the -OH group of alcohol attaches to a carbon atom with zero or one extra carbon atom. When the -OH group of an alcohol is linked to two more carbon atoms, a ketone is produced. There will be no C-H bond to replace it if the molecule contains three carbons connected to the carbon attached to the -OH, rendering it resistant to oxidation.

The formaldehyde vapor is unpleasant and odorless. Its chemical formula is HCHO. Formalin is an aqueous solution containing roughly 37% formaldehyde by weight that is used to provide formaldehyde. By enhancing protein coagulation and blocking a variety of biological processes that lead to tissue degradation, formaldehyde kills bacteria and other organisms. Formaldehyde is used to preserve tissue samples and embalm the deceased. It is also used to disinfect soil and other pollutants. Bakelite is a chemically and electrically stable polymer made from formaldehyde.

The most fundamental ketone is dimethyl ketone, generally known as acetone. Both 2-propanol oxidation and grain or molasses fermentation are used in commercial manufacturing. Acetone is a clear liquid with no color. It is used as a solvent for lacquers (including nail polish), plastics, varnishes, cellulose acetate, cellulose nitrate, acetylene, and the manufacture of pharmaceuticals and drugs, in addition to removing paint and varnish.

6.2.2. Carboxylic Acids (Alkanoic Acids)

When the C=O group combines with an OH, the resulting C(=O) OH structure is a brand-new functional group known as the carboxy group or carboxylic acid group, which is present in carboxylic acids. Once more, we demonstrate its overall structure, in which the R group may be H or a C directly attached to C=O. Carboxylic acids are analogous to aldehydes in that an O has been inserted between the H and C=O to create an OH group. This analogy is beneficial given that carboxylic acids are formed through the oxidation of aldehydes, as stated previously. By substituting the carboxylic acid for the aldehyde's final "al" in their systematic names, aldehydes are frequently derived from the names of their precursor aldehydes. "Cycloalkane carboxylic acids" are an exception to this rule. Ions of the carboxylate form. In the reaction between bases and carboxylic acids, carboxylate ions are produced (Chang, 2012). We identify carboxylate ions by systematically removing "-ic acid" and adding "-ate" to the names of their source carboxylic acids. When referring to simple acids with substituents by their common names, Greek letters rather than numbers are used to indicate the locations of the substituents. These Greek letters represent the position of the carbon atom that contains the substituent in the C=O group. Greek letters are not used in systematic nomenclature, but they can be used to explain how a substituent is connected to the C=O group. Frequently, common names for carboxylate ions are derived from common names for carboxylic acids by removing "ic acid" and substituting "ate." The longest hydrocarbon chain containing the -COOH carbon is the progenitor chain for acyclic-saturated monocarboxylic acids. The final -e of the progenitor hydrocarbon name is then changed to -oic acid. The progenitor carbon chain is numbered beginning with the -COOH carbon and ending with the -COOH carbon. Following the parent's name is an alphabetical enumeration of the substitutes' names and locations (Wang, 2015).

Notably, in the presence of -COOH groups, all other functional groups, including -CHO, -CO-, -OH, -OR, -NH_2, and halogens, are classified as substituents because they are ascribed less importance than the -COOH group. The name of the unsaturated carboxylic acid is derived from its alkene or alkyne progenitor. Multiple bonds are denoted by an additional locant appended to the alkene or alkyne progenitor. The parent alkene is prefixed with cis/trans or E/Z to reflect the stereochemistry surrounding the double bond. Similar nomenclature rules apply to acyclic dicarboxylic acids, with the exception that the parent hydrocarbon's terminal -e is retained and the suffix -dioic acid is added.

When perceiving one of the -COOH groups as a substituent, it is possible to generate names for compounds with more than two -COOH groups. The following carboxylic acid, for instance, adds the prefix carboxymethyl- to the parent's name and considers the -CH_2COOH group to be a substituent. Parent ring names of cyclic carboxylic acids with -COOH groups directly affixed to a cycloalkane or aromatic ring are suffixed with -carboxylic acid. In contrast to acyclic carboxylic acids, the -COOH-linked ring carbon is counted as carbon-1. In this nomenclature, benzoic acid is referred to as benzene-carboxylic acid.

When referring to substituted aromatic acids with one carboxyl group as benzoic acid derivatives, the locators are used to designate the position of the substituents relative to the carboxyl group on the benzene ring.

According to tradition, C-1 is the carbon to which the carboxyl group is attached. The systematic name of an aliphatic dicarboxylic acid with an open chain is formed by adding the suffix -dioic acid to the name of the parent hydrocarbon, alkanedioic acid. Attaching the suffix -dicarboxylic acid to the name of the parent cycloalkane or arene generates the systematic name of a cyclic aliphatic or aromatic dicarboxylic acid, i.e., cycloalkane-dicarboxylic acid or arene-dicarboxylic acid. To differentiate between isomers, the coordinates of the two carboxyl groups are provided with the minimum possible address numbers (Wang, 2015).

Commonly, carboxylic acids with six or fewer carbon atoms are referred to by their common names. Historically, these names were assigned by early chemists to characterize a property of the substance, frequently its origin. Formic acid, whose name derives from the Latin word for "ant," Formica, is found in ants, bees, and other venomous invertebrates. Acetum is the Latin word for vinegar, and acetic acid, which is prevalent in vinegar, derives its name from this word. Derived from the Greek terms pro ("the first") and pion ("fat"), propionic acid is the smallest acid that possesses some of the characteristics of larger fatty acids. Butyric acid is present in rancid butter; the Latin term for "butter" is butyrum. Valeric acid is found in valerian plant roots. If you scent both a goat and caproic acid, you will find that their odors are identical. Caproic acid is detectable in goat's milk. The Latin word "caper" means "goat."

A number identifies the position of a substituent in IUPAC nomenclature. A carboxylic acid's carbonyl carbon is always carbon-1. In traditional nomenclature, the carbonyl carbon is not designated, and the position of a substituent is denoted by a lowercase Greek letter. The carbon immediately adjacent to the carbonyl carbon is referred to as the Alpha Carbon, followed by the Beta Carbon, etc. Acyl halides are compounds of carboxylic acid in which the OH group has been replaced by a halogen atom. Acyl chlorides and acyl bromides are the two most prevalent acyl halides. The IUPAC name for these compounds is alkanoyl halide. When naming acyl halides, the term "ic acid" is supplanted with "yl chloride" (or "yl bromide"). When an acid's name concludes with "carboxylic acid," "carboxylic acid" is changed to "carbonyl chloride" (or "bromide"). When two carboxylic acid molecules shed their water, an acid anhydride is formed. The term "anhydride" signifies "without water." If the two molecules of carboxylic acid that comprise the anhydride are identical, the acid anhydride is symmetrical. If the molecules of the two carboxylic acids differ, the anhydride is referred to as a mixed anhydride. The name of the acid is used to designate symmetrical anhydrides, but "acid" is changed to "anhydride." When identifying mixed anhydrides, the names of the two acids should be listed in alphabetical order, followed by "anhydride" (Figure 6.5).

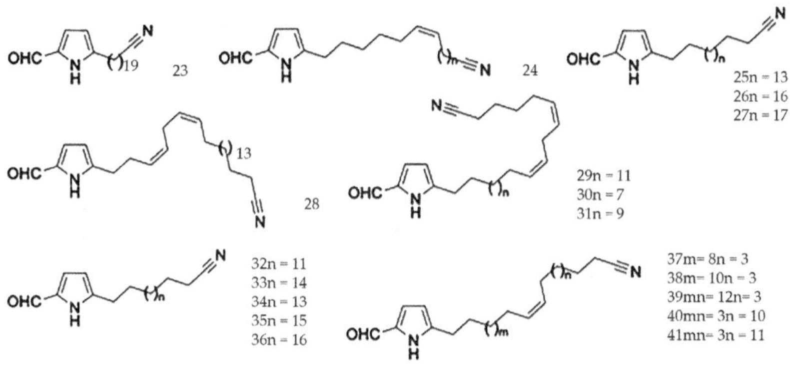

Figure 6.5. Natural compounds containing the nitrile functional group.

Source: *Scotti, & Barlow, (2022).*

A carboxylic acid in which the OH group has been replaced by an OR group is an ester. The two-word generic designation assigned by the IUPAC is alkylalkanotae. When identifying an ester, the group linked to the carboxyl oxygen is listed first, followed by the name of the acid with "ic acid" replacing "ate." Carboxylic acid salts have similar names. The name of the cation is given first, followed by the name of the acid, with "ic acid" once again supplanted by "ate." Lactones are used to refer to cyclic esters. They are known scientifically as "2-oxacycloalkanones." Their common names are derived from the common name of the carboxylic acid, which indicates the number of carbons in the chain, and a Greek letter representing the carbon to which the carboxyl oxygen is affixed. The carboxyl oxygen is located on the beta carbon in four-membered ring lactones, the gamma carbon in five-membered rings, and the delta carbon in six-membered rings (Chang, 2012).

The OH group of carboxylic acid is replaced by an NH_2, NHR, or NR_2 group in an amide. The IUPAC name for these compounds is N-alkylalkanamide. Amides are named by replacing "amide" in the name of the acid with "oic acid" or "ic acid." In the definition of acids with the suffix "carboxylic acid," "ylic acid" is replaced by "amide." If there is a bond between a substituent and nitrogen, the name of the substituent is listed first, followed by the name of the amide (in alphabetical order if there are multiple). Each substituent name is followed by a capital N to signify that it forms a covalent bond with a nitrogen atom. Lactams are the name for cyclic amides. Their names are spelled identically to lactones. In systematic nomenclature, they are referred to as "2-azacycloalkanones" (where "aza" represents the nitrogen atom). The common designation of carboxylic acid signifies the length of the carbon chain, and a Greek letter identifies the carbon atom to which the nitrogen is bound (Wang, 2015).

Due to the presence of acetic acid, a carboxylic acid, vinegar has a distinctive odor. Esters, molecules formed when a carboxylic acid reacts with alcohol, are responsible for the aroma of mature bananas and a variety of other fruits. Since esters lack hydrogen bonds between their molecules, their vapor pressures are lower than those of the alcohols and carboxylic acids from which they are derived.

Both carboxylic acids and esters contain a carbonyl group with a single bond connecting a second oxygen atom to the carbon atom. The second oxygen atom in a carboxylic acid also forms a bond with a hydrogen atom. The second oxygen atom in an ester forms a covalent bond with carbon. Carboxylic acid and ester names are derived using nomenclature principles comparable to those used for inorganic acids and salts and include prefixes indicating the lengths of carbon chains in the molecules.

Carboxylic acids are weak acids, meaning they are not completely ionized in water (for more information, see the chapter on acids and bases). Approximately 1% of the water molecules dissolved in a carboxylic acid are ionized at any given time. The remaining molecules are still connected in the solution.

Formic acid, HCO_2H, has been recognized as the most fundamental carboxylic acid since 1670. The name is derived from the Latin word formicus, which means "ant" It contributes to the discomfort and irritation caused by ant and wasp attacks and is responsible for the distinctive odor that occasionally emanates from ant nests (Chang, 2012).

Vinegar contains 3–6% acetic acid (CH_3COOH). Apple juice can be fermented without the addition of oxygen to produce cider vinegar. The yeast organisms in liquid are responsible for fermentation. The juice's sugar is fermented to produce ethanol, which is then followed by the production of acetic acid. Pure acetic acid usage results in a pungent odor and lesions. It is necessary for the production of cellulose acetate, which is used to dissolve a variety of organic and inorganic substances. It is also an essential ingredient in the manufacture of rayon, a ubiquitous synthetic fabric. Numerous florals, fragrances, and mature fruits have alluring aromas and flavors due to the presence of esters. Among the most important natural esters are fats (such as lard, tallow, and butter) and oils (such as linseed, cottonseed, and olive oils), which are esters of trihydroxyl alcohol glycerine, $C_3H_5(OH)_3$, with large carboxylic acids, such as palmitic acid. Double-bonded oleic acid is an unsaturated C=CC=C acid. The fatty acids are palmitic and stearic and lack double and triple bonds (Wang, 2015).

Amines are compounds in which one or more of the hydrogen atoms present in ammonia have been substituted by alkyl groups. Smaller amines emanate a distinctive salty odor. There are three classes of amines: primary, secondary, and tertiary. The classification is determined by the number of alkyl groups bound to nitrogen. Primary amines have one alkyl group attached to nitrogen, secondary amines have two, and tertiary amines have three. Remember that the number of alkyl groups attached to the nitrogen determines the classification of an amine as primary, secondary, or tertiary. In contrast, the classification of an alkyl halide or alcohol is determined by the number of alkyl groups attached to the carbon to which the halogen or OH is coupled.

The identities of the alkyl groups linked to nitrogen are included in the common designation of an amine, which is followed by the word "amine." In contrast to the common names of alcohols, ethers, and alkyl halides, in which "alcohol," "ether," and "halide" are written as discrete syllables, the phrase is written as a single word. The IUPAC system uses the term "alkylalkanamine" to define amine. In the IUPAC system, the amine functional group is denoted by a suffix. As with alcohols, the "amine" replaces the "e" at the end of the progenitor hydrocarbon's name. A number is ascribed to the carbon atom to which nitrogen is covalently bonded. The number may precede "amine" based on the name of the progenitor hydrocarbon. Any alkyl group linked to nitrogen is preceded by an "N" (in italics) to denote that it is linked to nitrogen as opposed to carbon (Chang, 2012).

The substituents are enumerated alphabetically, irrespective of whether they are connected to the parent hydrocarbon or nitrogen, and each is designated a number or "N" afterward. The chain is numbered in reverse to provide the lowest possible number for the functional group suffix.

As quaternary ammonium ions, nitrogen molecules with four bonded alkyl groups have a formal positive charge. The words "ammonium" and the name of the counterion are combined with the alphabetical names of the alkyl groups to form their names.

6.2.3. Cyclic Ketones and Diketones

The carbonyl group in cyclic ketones is allotted position 1, and this number is not included in the name unless there are many carbonyl groups. To provide alternatives with the lowest feasible location numbers, the remaining segments of the ring are numbered. Remember that the name of the parent chain is prefixed with cyclo to denote that it is a ring. As with other ketones, the -e ending is replaced with -one to denote the presence of a ketone.

Carbonyls must be assigned location numbers in cycloalkanes containing two ketones. In addition, the suffix -dione is appended instead of the final -e being deleted (Wang, 2015).

6.3. NITRILES

The rules of nomenclature are complicated. We will only attempt to identify relatively basic molecules. Consider the following examples of compounds with ester groups. Primary amides are dehydrated to form nitriles. If the CN is the primary function in the main chain, the suffix "nitrile" is added to its name. It's where the tallying begins. If the CN serves a secondary purpose, it is categorized as a "cyano" substituent with its localizing number.

Nitriles are substances with a functional group. Since nitriles react with water to form carboxylic acids, they are classified as derivatives of carboxylic acids. In systematic nomenclature, nitriles are identified by adding "nitrile" to the name of the progenitor alkane, Alkanenitrile. Note that the triple-bonded carbon in the nitrile group is factored into the length of the longest continuous carbon chain. Common nomenclature defines nitriles by substituting "onitrile" for "ic acid" in the name of carboxylic acid. They are also known as alkyl cyanides to denote the alkyl group that is affixed to the cyanide group (Chang, 2012).

The functional category in organic chemistry, CN, also known as nitriles or cyanides, is a functional group in which the carbon and nitrogen atoms form a triple bond. Carbon and nitrogen atoms are also sp-hybridized, resulting in the generic formula RCN, where R represents the hydrocarbon group or component. Because of the presence of electrons on the nitrogen atom, the structure of nitriles was strikingly similar to that of alkynes. "nitrile" is used at the end of the primary chain of aliphatic and aromatic nitriles, while "cyano" is used as a substituent. Carbonitrile is added to cycloalkanes as a suffix (Wang, 2015).w

CHAPTER

7 Nomenclature of Aldehydes & Ketones

Contents

7.1. INTRODUCTION

Organic chemical compounds are named using the International Union of Pure and Applied Chemistry (IUPAC) nomenclature. Each organic compound needs a name and structural formula. Inorganic compounds are the IUPAC nomenclature. Except for chemical designation, the stringent IUPAC naming standards are usually ignored in everyday communication to avoid long and boring names. Ethanol is simpler to IUPAC than ethyl alcohol. They're easier than non-systematic molecular names. The normal phrase is shorter and simpler. Chemical sources may inspire non-systematic names. Structured names might be hard to understand. Chemical prefixes, suffixes, and interfixes characterize functional groups. Designating an organic compound involves several steps:

- Hydrocarbon chain progenitor. This chain must obey priority rules:
- The best functional group suffix choices. Primary functional groupings require suffixes.
- Use the most crucial function. Alliance limit. Maximize it. It should have the most prefixes. Possible single bonds. Identifying progenitor functional groups. Recognizing sidechains. Carbon side chains branch. Use ionic prefixes to identify residual functional groups, such as hydroxy for –OH (Ma, 2019)
- Alphabetize functional groupings and side chains. Words without prefixes are alphabetized. Ethyl precedes dihydroxy and dimethyl. Both times, "di" is ignored. Combine side chains and secondary functional groups.
- Bonds two or three strands. Sixth-chain numismatics. After numbering left to right and back, prioritize the numbers using these criteria. The fewest suffix functional group locants exist. Substituent-carbon locants exist. Minimum multiple bond locants. Minimum prefix locants.

Substitutions and linkages. A prefix is added to indicate the number of substituents or double bonds (di – 2, tri – 3, tetra – 4, then as for the number of carbons below with 'a' added). Increasing numbers must precede the auxiliary chain's name. Two identical alpha-carbon side chains are written twice. Aldehydes and ketones exist with several functional groups. Vanillin, Citra, helminthosporal, carvone, and camphor come from microorganisms and plants. Humans and animals have muscone, testosterone, progesterone, and cortisol. Aldehydes and ketones vary chemically. Oxidation produces carboxylic acids, making aldehydes more reactive than ketones. Ketones react less than aldehydes because they lack hydrogen. Potent oxidizers oxidize ketones at high temperatures. Ketones and aldehydes are IUPAC. Acids name oxidized aldehydes (Suslick, 2017) (Figure 7.1).

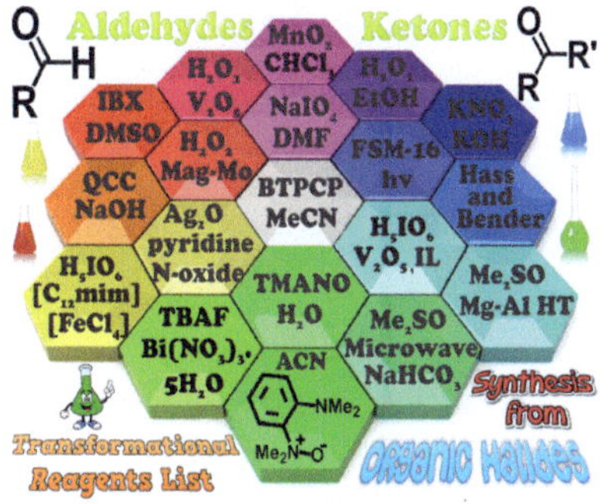

Figure 7.1. Aldehydes and ketones are important classes of chemicals that serve as important precursors for the synthesis of a library of compounds.

Source: *Faisal, Saeed, Shahzad, et al., (2020).*

Aldehydes and ketones react similarly due to their carbonyl functional group. Similar-molecular-weight alkanes and aldehydes have different melting and boiling temperatures. Aldehydes and ketones' polar C=O double bonds cause dipole-dipole forces. They boil lower than related alcohols. Alcohols produce hydrogen bonds due to their oxygen-hydrogen bonds. Hydrogen bonds need more energy than dipole-dipole forces (Ma, 2019).

Formaldehyde has three bonds: a carbon or hydrogen bond, two oxygen connections, and one hydrogen bond. Centered Sp^2-hybridized carbon. Polar aldehyde. Aldehydes are rare in amino acids, nucleic acids, and lipids due to the reactivity of the formyl group, although cinnamonaldehyde, cilantro, and vanillin contribute to essential oil scents. Aldehydes make most carbs. Hemiacetal aldoses hide aldehydes. Aqueous glucose aldehyde is negligible.

The order of the functional groups of a compound determines whether they are prefixed or suffixed. Below, common categories are displayed in descending order. Prefixes are used by everyone except the highest priority category.

Only the suffixes (-en and -yn) with other suffixes are used with double and triple bonds. Without modifiers like di-, tri-, etc., the alphabetical sequence of chlorofluoromethane is followed. The position number of a functional group with several prefixes or suffixes is ethane-1,2-diol, not ethane-2,1-diol. Its formula is $CH_3CH(CH_3)$ $CH_2NH(CH_3)$ for N,2-dimethylpropanamine. The N position is shown before the number "1" for amides and amines.

7.2. NOMENCLATURE OF ALDEHYDES

The IUPAC nomenclature system is the foundation for the first approach, also known as systematic nomenclature.

- In this method, the progenitor alkane is the carbon chain with the longest carbonyl group. • The aldehyde is revealed when the suffix -e is changed to -al (Alkanal).
- There is no need to designate the carbonyl group of an aldehyde with a number because it can only be found at the end of the progenitor chain, which is Carbon 1. The practice of designating aldehydes after the common name of the corresponding carboxylic acid is known as common nomenclature. Aldehydes have the same name, but the suffix -ic acid is used instead of aldehyde (Suslick, 2017) (Figure 7.2).

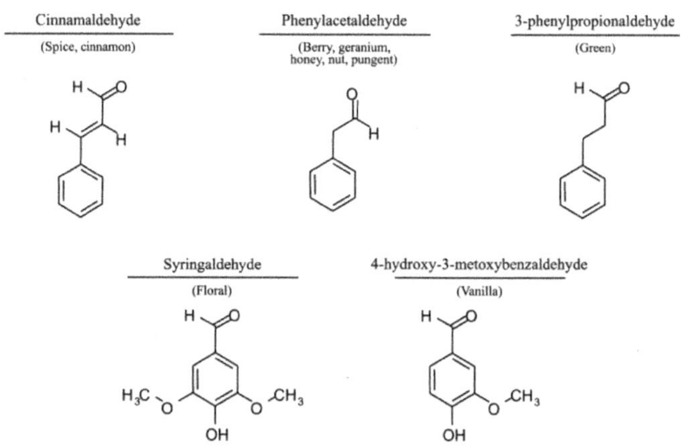

Figure 7.2. Phenolic aldehydes used in the experimental design.

Source: *Escott, et al., (2018); Scientific Figure on ResearchGate.*

7.2.1. Properties of Aldehydes

The presence of a carbonyl group is the sole structural distinction between aldehydes and hydrocarbons. The differences between physical and chemical properties are due to the carbonyl group. Naturally, the carbonyl group is polar. The C=O bond's electrons are attracted to oxygen rather than carbon. Carbon has a slight positive charge, while oxygen has a slight negative charge (Ma, 2019). Typically, the Greek symbol delta denotes a fractional charge and is used to denote the polarity of a carbonyl group. The polarity of the carbonyl group affects important physical properties, including melting temperature, boiling point, and solubility. Because polar and nonpolar molecules are more attracted to one another than nonpolar molecules are to polar molecules, separating them requires more energy and a higher temperature, which is required for compounds to dissolve or simmer. Normal conditions render formaldehyde (HCOH) a gas, whereas acetaldehyde (CH_3COH) boils at ambient temperature. Except for those with a high molecular weight, all other aldehydes are liquids under normal conditions. Numerous aldehydes possess appealing aromas (Suslick, 2017).

7.2.2. Uses of Aldehydes

- Different forms of plastics, including Bakelite and laminate tabletop material, are created by the fusion of several kinds of molecules. Formica Aldehydes may undergo a wide range of chemical processes, including polymerization.
- Vanillin is the substance that gives vanilla extract its pleasant aroma.
- The aldehyde molecule benzaldehyde gives almonds their distinctive scent. Cinnamonaldehyde may be present in cinnamon bark. By developing Chanel No. 5 in 1921, Russian perfumer Ernest Beaux is credited with popularizing the use of aromatic aldehydes in the industry. Aldehydes have since been used in the creation of fragrances and other fragrant goods.
- Retinal, an aldehyde form of vitamin A, and pyridoxal phosphate, a form of vitamin B6, are essential for human eyesight. Glucose and other lowering carbs, as well as several other natural and artificial hormones, make up aldehydes.
- Aldehydes are used in the manufacture of polymers, acids, and medicines as solvents and essential building blocks. Industries in addition, formaldehyde is used as a preservative, fungicide, and germicide (Ma, 2019).
- Aldehydes are used in the production of many different goods, including synthetic resins and pigments. In the production of Bakelite, Melamine, and sticky polymers, formaldehyde is used.

When the body breaks down fat for energy when there is not enough glucose present, the liver produces ketones, which are organic molecules. Similar to aldehydes, ketones have distinctive smells.

- A key concept in the study of organic chemistry is the functional group indicated by the formula RC(=O)R,' where R and R' stand for different numbers of carbon-containing substituents. The distinctive smell of the Himalayan musk deer is attributable to a particular functional group. Acetone, which has methylated R and R' groups, is regarded as the archetypal ketone. The parent alkane molecules from which it is produced are given the designation "ketone." The ending -e has been changed, and -one has taken its place.

Ketones may be recognized by their substituent groups' systematic nomenclature, which is alphabetically sorted and includes the word "ketone" after each group. The names of some frequently occurring ketones are used to identify them.

The organic substance with the chemical formula C_3H_6O, often known as acetone in scientific literature, is an organic compound with the common name propanone. In the current study, 18 ketones' olfactory properties are examined. Aldehydes and ketones, except low-carbon aldehydes, typically have a pleasant scent. With increasing molecular size, the olfactory characteristics of aldehyde and ketone molecules tend to become more aromatic and less pungent (Suslick, 2017).

The functional group in question consists of carbon-oxygen double bond. In a carbon-oxygen bond, electrons are pulled toward the oxygen atom because oxygen has a higher electronegativity than carbon. Positive and negative charges are dispersed from the elements carbon and oxygen. Political issues have increasingly affected the topic matter (Ma, 2019).

When a water molecule interacts with the carbonyl group's oxygen atom to create a connection, hydrogen bonding takes place.

Due to the absence of hydrogen atoms that are directly bonded to the carbonyl oxygen atom, aldehydes and ketones do not show intermolecular hydrogen bonding. The ability of the carbonyl oxygen to function as a hydrogen bond acceptor gives rise to the possibility of hydrogen bonding with water molecules. Small aldehydes and ketones are soluble in water due to the existence of hydrogen bonding.

7.2.3. Summary of Aldehyde Nomenclature Rules

- The term originates from the alkane chains present in precursor molecules of aldehydes. The final letter "e" is elided and subsequently replaced with the suffix "-al."
- Position number one is assigned to the aldehyde functional group however, it is not explicitly indicated in its nomenclature.
- To ascertain the conventional nomenclature of an aldehyde, it is necessary to affix the suffix -aldehyde to the common name of the parent chain. Greek symbols are utilized to denote substitute roles.
- Upon attachment of the -CHO functional group to a ring, the -carbaldehyde suffix is appended.

7.3. NAMING KETONES

Ketones in organic chemistry have the chemical formula RC(=O)R,' where R and R' represent a variable number of carbon-containing substituents. Ketones have a carbonyl group (C(=O)), a carbon-oxygen double bond. Acetone, $(CH_3)_2CO$, is the prototypical ketone. Biology and industry require ketones. Ketones, testosterone, and acetone are examples. Ketones are named per IUPAC guidelines. Adding -anone to the parent alkane's name creates a ketone. Conventional, non-systematic names for ketones like acetone and benzophenone are still widely used despite the widespread usage of number designations to represent the carbonyl group. Some introductory chemistry textbooks use "2-propanone" or "propan-2-one" instead of "acetone" to describe the basic ketone ($CH_3C(=O)CH_3$), however, the latter informal formulations are commonly acknowledged as IUPAC nomenclature (Suslick, 2017). IUPAC terminology uses -one for ketones. Usually identified by a number, a ketone's carbonyl functionality can be located anywhere along its chain or ring. Starting with the terminal nearest the carbonyl group, the chain numbering method is applied. For basic ketones like propanone and phenylethanone, locator numbers are unnecessary since there is only one carbonyl functionality site. Ketones are named from the two alkyl groups connected to the carbonyl group. Add "-ketone" to these names. Alkyl group nomenclature is alphabetical. First, the two alkyl groups related to the carbonyl group are named, followed by "ketone" as a distinct word. As shown by methyl ethyl ketone, alkyl group nomenclature has increased in complexity. IUPAC nomenclature rules identify alkyl groups like ethyl methyl ketone alphabetically. Organic chemistry uses "di-" to indicate two identical alkyl groups. Greek characters indicate group configurations with the -carbon atom adjacent to the carbonyl group (Ma, 2019).

IUPAC nomenclature uses "oxo" to prefix a secondary ketone's oxo group (=O). Prefixes abound. The ketone functional moiety, important in biological sciences, is called "keto" in chemistry. Ketone carbon atoms are "sp^2-hybridized" electrically and chemically. Ketones exhibit trigonal planarity around the ketonic carbon atom and bond angles of roughly 120° between C-C-O and C-C-C bonds. Ketones are carbon-based having a carbonyl moiety (C=O) linked to two carbon atoms, unlike aldehydes. Carbon chains finish in carbonyl groups. Carboxylic acids, esters, and amides are chemically separate from ketones. Oxygen's higher electronegativity than carbon makes the carbonyl group polar. Ketones are electrophilic to carbon and nucleophilic to oxygen. Due to hydrogen bonding, ketones are more water-soluble than methylenes. Ketones bind hydrogen. Ketones cannot form intramolecular hydrogen bonds; hence they seldom donate hydrogen bonds. Ketones are more volatile than alcohols and carboxylic acids of equivalent molecular weight because they cannot act as hydrogen-bond donors and acceptors, which prevents self-association. Ketones in solvents and perfumes are linked to their characteristics (Suslick, 2017).

7.3.1. Classes of Ketones

Ketones are categorized according to the substituents they contain. Based on the equivalence of the two organic substituents affixed to the carbonyl nucleus, ketones are categorized as symmetrical or asymmetrical derivatives. Examples of symmetrical ketones are acetone and benzophenone $(C_6H_5)_2CO$. The asymmetric ketone acetophenone, with the formula $C_6H_5O(O)CH_3$, has the molecular formula $C_6H_5O(O)CH_3$ (Figure 7.3).

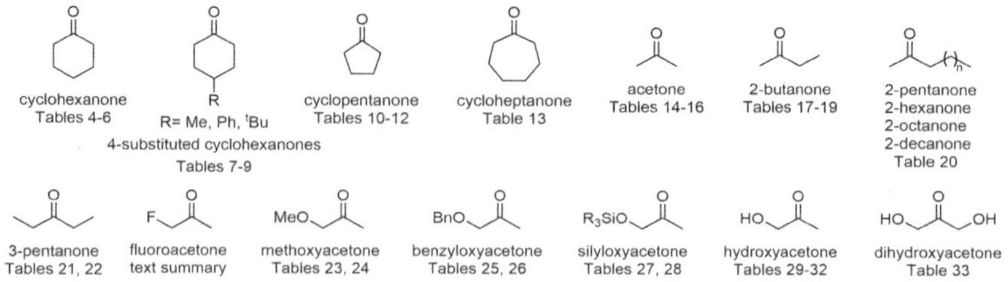

Figure 7.3. A summary of ketones.

Source: *Nugent, et al., (2021); Scientific Figure on ResearchGate.*

There are numerous forms of diketones, some of which have distinctive properties. The most fundamental is diacetyl ($CH_3C(O)C(O)CH_3$), which was formerly used to butter-flavor popcorn. Since it is most commonly found as the monoenol $CH_3C(O)CH=C(OH)CH_3$, the name acetylacetone (pentane-2,4-dione) is a misspelling. Its enolate is a typical ligand in the field of coordination chemistry (Ma, 2019).

- ***Unsaturated Ketones*:** The term "unsaturated ketones" refers to ketones composed of alkene and alkyne units. The most frequently employed chemical in this category is methyl vinyl ketone, or $CH_3C(O)CH=CH_2$, which is essential to the Robinson annulation procedure. To avoid any confusion, a ketone is a location that can be hydrogenated and is unsaturated.

7.3.2. The Ketone Cycle

The formula for the simplest class is $(CH_2)_nCO$, where n ranges from 2 for cyclopropanone ($(CH_2)_2CO$) to the tens. There exist more intricate derivatives. Cyclohexanone ($(CH_2)_5CO$), a symmetrical cyclic ketone, is an essential step in the manufacturing of nylon. Isophorone is an unsaturated, asymmetrical ketone derived from acetone that serves as a building element for other polymers. Muscone, also known as 3-methylpentadecanone, is a pheromone. Another cyclic ketone with the formula $(CH_2)_3CO$ is cyclobutanone. Through keto-enol tautomerization, ketones containing at least one alpha-hydrogen are converted into enols. Both acids and bases stimulate tautomer production. The keto form is frequently more stable than the enol form. This equilibrium permits the hydration of alkynes to yield ketones (Suslick, 2017).

7.3.3. Acid/Base Properties of Ketones

Ketones (pKa 20) have more acidic C-H bonds near the carbonyl than alkanes (pKa 50). The resonance stability of the enolate ion following deprotonation causes this fluctuation. Enolization reactions of ketones and other carbonyl compounds are dependent upon the relative acidity of the -hydrogen. Ketones and other carbonyl compounds can react as nucleophiles at that position with either a stoichiometric or catalytic base due to the acidity of the -hydrogen. The less-substituted kinetic enolate is produced selectively with strong bases such as lithium diisopropylamide (LDA, pKa of conjugate acid 36), whereas the more-substituted thermodynamic enolate is produced under equilibration conditions (higher temperature, base added to ketone, weak or insoluble bases such as CH_3CH_2ONa in CH_3CH_2OH, or NaH) (Ma, 2019).

In the presence of Brnsted acids, the carbonyl oxygen of ketones is protonated, transforming them into weak bases. Protonated ketones, also referred to as ketonium ions, are potent acids with pKa values between -5 and -7. Even though acids encountered in organic chemistry are rarely strong enough to completely protonate ketones, the formation of equilibrium concentrations of protonated ketones is a crucial step in the processes of numerous common chemical reactions, such as the synthesis of an acetal. Despite the extremely unfavorable protonation equilibrium constant (Keq 1010), weak acids with a pKa of 5.2, such as the pyridinium cation (present in pyridinium tosylate), can function as catalysts under these conditions.

7.3.4. Characteristics of Ketones

Aldehydes are simpler to oxidize than ketones because they have an affixed hydrogen atom to their carbonyl group. Because they lack a hydrogen atom attached to the carbonyl group, ketones are more resistant to oxidation. They can only be oxidized by strong oxidizing chemicals capable of disrupting carbon-carbon bonds. Around 1700 cm1, ketones and aldehydes have significant infrared absorption. The precise apex placement is determined by the substituents.1H NMR spectroscopy is sometimes insufficient to demonstrate the existence of a ketone, even though, depending on the structure, 13C NMR spectra exhibit signals substantially below 200 ppm. Due to the absence of nuclear Overhauser effects, these signals are typically feeble. Because aldehydes and ketones undergo similar chemical changes, numerous resonance experiments are used to distinguish between them (Suslick, 2017).

***Organic Quality Assessments*:** In the test conducted by Brady, ketones react with 2, 4-dinitrophenylhydrazine to form the corresponding hydrazone. To isolate ketones from aldehydes, either Tollens' reagent or Fehling's solution must yield a negative result. The iodoform test reveals the presence of methyl ketones. In the presence of diluted sodium hydroxide and m-dinitrobenzene, ketones produce a violet hue.

***Synthesis*:** In both commercial and academic laboratories, ketones can be produced in several methods. The biochemistry section that follows contains additional details on how organisms produce ketones. The oxidation of hydrocarbons, typically with oxygen, is without a doubt the most essential industrial process. For instance, one billion kilograms of cyclohexanone are produced annually through the aerobic oxidation of cyclohexane. Cumene is oxidized by oxygen to produce acetone. Secondary alcohol oxidation is a prevalent method for producing ketones for applications in specialized or small-scale organic synthesis: Potassium permanganate and a Cr(VI) compound are examples of potent oxidants (the source of "O" in the aforementioned process). Dess-Martin periodinane or Moffatt-Swern procedures are utilized for milder instances. Other strategies that have been developed include: The geminal halide is hydrolyzed in step 9. Alkynes are soluble in water. For these enol-mediated reactions, an acid and mercury(II) sulfate ($HgSO_4$) are required. Additional enol-keto tautomerization leads to the formation of ketone. Except for acetylene hydration, which yields acetaldehyde, this reaction always yields a ketone, even in the presence of a terminal alkyne (Ma, 2019).

- Utilizing the stoichiometric organometallic reagents of Weinreb Amides.
- Aromatic ketones can be generated through the Friedel-Crafts acylation, the accompanying Houben-Hoesch reaction, and the Fries rearrangement.
- Depending on the pattern of alkene substitution, ozonolysis and related dihydroxylation/oxidative sequences cleave alkenes into aldehydes or ketones.
- The Kornblum-DeLaMarre rearrangement uses base and peroxides to generate ketones.
- The Ruzicka cyclization method converts dicarboxylic acids into cyclical ketones
- By hydrolyzing salts of secondary nitro compounds, the Nef reaction produces cyclical ketones.
- In the Fukuyama coupling, a thioester and an organozinc molecule interact to create ketones.
- The Dakin-West reaction is an effective technique for producing specific methyl ketones from carboxylic acids. Combining an acid chloride with organocadmium or organocopper compounds accomplishes this. Nitriles and Grignard reagents can be combined to produce ketones, which are then hydrolyzed.
- Haloketones can be converted to ketones through the dehalogenation of halo ketones.
- Through carboxylic anhydride decarboxylation.

- During ketonic decarboxylation, carboxylic acids are converted into symmetrical ketones
- Amine oxidation by chloride of iron(III).
- Hydrolysis of secondary unsaturated amides, -keto acid esters, or -diketones. The 1,2-diol restructuring is catalyzed by acid (Suslick, 2017).
- When a potent amide base and an unanalyzable ketone react, the Haller-Bauer reaction occurs. In this typical benzophenone reaction, the tetrahedral intermediate discharges the phenyl anion, yielding benzamide and benzene as byproducts. Ketones are utilized in numerous organic processes. The most significant reactions are caused by the propensity for nucleophilic addition of the carbonyl carbon and the propensity for electrophilic addition of the enolates. Nucleophilic additions are arranged roughly in order of decreasing universality:
  - Geminal diols produced by hydration are not always detectable or present in significant quantities.
  - Utilizing an acetylide to produce -hydroxyalkyne. To create an imine, ammonia or a primary amine is used. To create enamine, a secondary amine is used.
  - Following an aqueous workup, create a tertiary alcohol using Grignard and organolithium reagents.
  - The hemiketal or its conjugate base is produced when an alkoxide or alcohol is reacted with a diol. This reaction is used to preserve the safety of ketones.
  - In the presence of sodium amide, the Haller-Bauer reaction breaks the C-C bond, producing the amide $RCONH_2$ and the alkane or arene R'H.
  - Creating carboxylic acids using potent oxidizers. Frequently, robust oxidizing agents and high temperatures are used to actively oxidize ketones. By cleaving carbon-carbon bonds, various carboxylic acids with fewer carbon atoms than the progenitor ketone are produced (Ma, 2019).
  - Electrophilic addition and contact with an electrophile produce a resonance-stabilized cation.
  - The Wittig reaction, which utilizes phosphonium yields to produce alkenes, produces resonance-stabilized cations.

Thiols can be used to create thioacetals. Hydrazones can be produced using hydrazine or 1-disubstituted hydrazine derivatives. An example of ketone reduction is the hydrolysis of a metal alkoxide salt in the presence of a metal hydride, which produces alcohol. The use of halogens to initiate an enol-mediated reaction that produces a haloketone. Heavy water Photochemical Norrish reaction fragmentation produces a -deuterated ketone. The Robinson-Gabriel synthesis converts 1,4-aminodiketones to oxazoles via dehydration. In the presence of sulfur and an amine, the Willgerodt reaction converts aryl-alkyl ketones into amides. To generate secondary alcohols, combine hydroxylamine with reducing agents. Utilizing peroxy acids, the Baeyer-Villiger oxidation produces esters. Ketones can be discovered everywhere in nature. The ketone ribulose-1,5-bisphosphate is involved in the production of organic compounds during photosynthesis. Numerous sweeteners contain ketones, also known as ketoses. Fructose, the most well-known ketose, primarily occurs as a cyclic hemiketal, which conceals the ketone functional group. Fatty acids are synthesized by utilizing ketones. Acetoacetate is an intermediate in the Krebs cycle, which converts glucose and carbohydrates into energy (Ma, 2019).

Medicine refers to acetone, acetoacetate, and beta-hydroxybutyrate (BHB) as ketone substances. Most vertebrates, including humans, produce ketone bodies from carbohydrates, fatty acids, and amino acids. There is an increase in ketone bodies in the blood after fasting, including a night of sleep (ketosis); there is an increase in ketone bodies in the blood and urine during starvation; in hypoglycemia from causes other than hyperinsulinism; in several inborn metabolic errors; and when ketoacidosis is deliberately induced (typically due to diabetes mellitus). Ketoacidosis is a characteristic of decompensated or uncontrolled type 1 diabetes, but ketosis or even ketoacidosis can occasionally occur in type 2 diabetes (Suslick, 2017).

7.3.5. Applications of Ketones

Massive quantities of ketones are produced by the pharmaceutical, polymer, and solvent industries. Methylethyl ketone, cyclohexanone, and acetone are the three most significant ketones in terms of size. Although they occur less frequently in biochemistry than in general organic chemistry, they are nonetheless common. During combustion, which is the uncontrolled oxidation of hydrocarbons, several compounds, including ketones, are produced. Even though it is challenging to generalize about the toxicity of such a large family of compounds, simple ketones are typically not exceedingly hazardous. This is one reason why they are utilized as solvents.

Unsaturated ketones, such as methyl vinyl ketone, which has an oral LD50 of 7 mg/kg, fail to meet this criterion (Ma, 2019).

7.3.6. Summary of Ketone Nomenclature Rules

- Ketones derive their name from the alkane compounds from which they are derived. The -e suffix has been changed to -one.
- Ketones are readily distinguished by their substituent groups, which are listed after the word "ketone"
- Some prevalent ketones have widely recognized names. For example, propanone is also known as acetone.

7.4. NAMING ALDEHYDES AND KETONES IN THE SAME MOLECULE

Aldehydes are generally referred to as "oxo" substituents, and they are used more frequently than ketones. Most compounds show a preference for a certain functional group, which is referred to as "substituents." Based on the assumption that dialdehydes functional groups are present at the parent chain's terminals, both carbonyls are bypassed. Ketones only require one component. In cyclic ketones, the carbonyl group occupies position 1. The remaining circuit is changed starting with the locations with the lowest numbers. The word "Cyclo" is a shortened version of the word "ring" as it relates to the basic nomenclature of a chain. The presence of a carbonyl group, indicated by the suffix -one, distinguishes ketones from other chemical molecules.

Carbonyl sites are required for cycloalkanes with two ketone functional groups. The prefix "-dione" is used in place of the prefix "-e." Aldehydes, ketones, and alcohols are given precedence in the IUPAC naming scheme for carbonyl compounds. The lowest location number and proper suffix are given to the carbonyl functional group in nomenclature. Alcohols have more molecules than hydroxyl (OH), which has only one. Frequently, hydroxyl groups are deprotonated. The carbonyl functional groups found in aldehydes, ketones, and alkenes are given priority by the IUPAC. The nomenclature designates the carbonyl functional group with the appropriate suffix and the lowest location number. Having a carbonyl structure, alkenes (Ma, 2019).

The biggest carbon chain, omitting the -ane ending, is given a prefix, followed by the -en suffix, to denote an alkene in the nomenclature of alkenes. The position number of the carbonyl group is also provided in the case of ketones. The molecule is then given the suffixes -one or -anal.

The carbonyl moiety, which is the most important component, should be positioned at the lowest location number. Alkenes should, wherever feasible, be labeled as cis/trans or E/Z.

Water may dissolve acetone, acetaldehyde, and formaldehyde. A substance's carbon chain length is positively connected with how soluble it is in water. When there are four carbon atoms for every oxygen atom, the greatest amount of solubility is attained. The solubility of aldehydes and ketones in organic solvents is a result of their lower density than water. The polarity of the carbonyl group is what causes aldehydes and ketones to have higher boiling points than alkanes and ethers. These compounds' lower boiling points can be due to the lack of hydrogen bonds that set them apart from related alcohols and carboxylic acids (Suslick, 2017).

It has been shown that a substance's water solubility rises as the length of its carbon chain does. The ratio of four carbon atoms to one oxygen atom results in the best solubility. Because they are less dense than water, organic solvents can dissolve aldehydes and ketones. Aromatic ketones are used in the creation of fragrances. Various plant extracts include essential oil ketones such as acetophenone, carvone, and jasmone.

The polarity of their carbonyl group can be used to explain why aldehydes and ketones have greater boiling points than alkanes and ethers. These compounds' lower boiling points can be due to the lack of hydrogen bonds that set them apart from comparable alcohols or carboxylic acids. Various natural sources contain essential oil ketones such as acetophenone, carvone, and jasmone.

These materials are used as main ingredients and solvents in the adhesives, paint, rubber, printing, and electroplating industries. The rubber and paint industries both use methyl isobutyl ketone. Aldehydes and ketones are necessary for human nutrition. The cortisol, testosterone, progesterone, and glucose hormones are the four in concern. The metabolic state known as "ketosis" is when the body uses ketone bodies as its main energy source as opposed to starting the state of ketosis. When fat is used as the body's major energy source instead of glucose, acetone, acetoacetate, and beta-hydroxybutyrate (BHB) are produced. This metabolic state is known as ketosis (Ma, 2019).

Proteins and lipids are prioritized above carbs in the diets associated with ketosis. The ketogenic diet has become more well-known as a weight loss method; despite being first created to treat infantile epilepsy. It is essential to remove the suffix "ek" from the matching parent alkane to produce an aldehyde. A functional group comprised of a carbon atom double-bonded to an oxygen atom (-C=O) distinguishes aldehydes from other organic compounds. Aldehydes must include a minimum of one hydrogen atom in order to exist. Organic molecules known as ketones are distinguished by the presence of a carbonyl group, which is made up of two carbon atoms. Learn common terms for ketones and aldehydes, such as acetaldehyde and formaldehyde. There is a decline in the use of standardized names. Dihydroxyacetone (DHA) is less common than 1,3-dihydroxypropanone-3-heptanone. Don't try to break the connection, please. Despite being categorized as a low-level carcinogen, formaldehyde, also known as methanal, is used to preserve biological material. Implementing optimum ventilation leads to optimal performance. Nail polish removers include acetone, a substance known to cause cancer. Uncontrolled diabetes slows down the metabolic rate. It's possible to characterize the breath's aroma as "fruity" or as resembling nail polish remover. The presence of ketones in the bloodstream is referred to as ketosis. It is striking how common ketoacidosis is, a disorder marked by noticeably high blood acidity levels. Self-tanners frequently contain the chemical dihydroxyacetone. The process results in the creation of melaninoids with a dark hue by interacting amino acids found in the epidermis. The formula was altered, and as a consequence, the color was improved to a more vivid orange. Due to its low capacity to absorb ultraviolet A (UV-A) rays between 320 and 400 nanometers, even sunscreen with a sun protection factor (SPF) of 3 has the potential to result in sunburn. Sort a collection with two distinctive features into categories. Starting at the terminal carbon atom of the molecule, which is given the lowest number, one should count backward to find the carbon atom of the carbonyl (C=O) group. Aldehydes would show the earliest arrival regardless of the sequence. The reactivity of carbonyl groups may be lessened by the presence of halogen or alkyl groups. Aldehydes and ketones can both be found in molecules. Diacetyl, sometimes referred to as glutaraldehyde, is a tiny amount naturally occurring result of yeast fermentation. The advantages of butter and Chardonnay consumption. Popcorn and margarine have been associated with buttery flavors for a very long time. Insufficiently ventilated popcorn companies have reported rare cases of bronchiolitis obliterans. Even though consumers are exposed to diacetyl at far lower levels than those involved in the production, several popcorn firms have changed. Three active food efforts are currently free of diacetyl. Most toxicologists believe that diacetyl, which makes up the bulk of fake butter, is safe (Suslick, 2017).

Non-autoclavable medical equipment was sterilized with glutaraldehyde (1,5-pentane diol) before being used to fix microscopy specimens. Similar to formaldehyde, the substance's aldehyde groups have virucidal and germicidal effects. Due to the significant skin and respiratory discomfort associated with sterilizing treatments, their use has decreased as more practical alternatives have been created. Four more ketones and aldehydes are thought to be beneficial. Organic substances like ketones and aldehydes are frequently used as flavoring agents. Almonds have a distinctive taste that comes from benzaldehyde. Vanillin is a component of vanilla tastes. Cinnamaldehyde is thought to be the source of cinnamon's aroma. Petroleum-derived cinnamon is available. The artificial cinnamon taste is added by the presence of formaldehyde-derived chemicals. Turpentine, oil, and plants from Southeast Asia are used to make camphor oil. On the skin, Vicks VapoRub offers a cooling and calming effect and eliminates obstacles. The presence of carvone 6 R-carvone, which lends a characteristic flavor reminiscent of caraway seeds, is credited with giving rye bread its flavor profile. S-carvone is the flavoring compound found in spearmint (Ma, 2019).

What makes these molecules unique? As a result, isomerism is present in the class. Enantiomers, or molecules with mirror-image symmetry, may be distinguished using the sensory senses of taste and smell. What functional groups are found in glucose, fructose, ketones, and polyhydroxy aldehydes? Fructose? Solubility is the last factor to take into account. Due to the C=O group's weaker polarity than the O-H group's, it is hypothesized that water can dissolve up to four CH_2 units. The purpose of this study is to determine how soluble formalin (1,1 methanediol) is in water. When preserving dead organisms, formalin is used as a preservation agent. When formaldehyde is present in its hydrate form with a concentration of 99.9%, the equilibrium process moves to the right. The oxygen atom in water transfers its electrons into the carbonyl group upon absorption by the pi electrons of the C=O bond. The hydrogen ion and the electronegative oxygen atom share electrons in the chemical reaction C=O. The arrow of the chemical-electron process is back. The presence of an oxygen atom, which has three electron pairs but only needs two for a stable configuration, causes the oxonium ions to have a positive charge. The oxygen atom causes the carbonyl's partial positive charge to rise by pushing its electron pairs closer together. The migration of the water molecule O in the direction of the carbonyl group C, which has a high electronegativity, is facilitated by the concerted electron rearrangement process. All four bonds are kept intact thanks to the two electrons that C=O and water share. The oxygen atom of water gains a total of three electron pairs and eight positive charges as a result of sharing electrons with carbon. By giving up one of its shared hydrogen ions, the oxygen atom within the water molecule undergoes deprotonation, regaining its neutral charge. By using an arrow, shared electrons between atoms are shown to be returning to the oxygen atom. The proton that was lost during the original chemical reaction is replaced to restore neutrality. Proton replenishment is necessary for the catalytic reaction. The early reactions and hydration process will eventually attain an equilibrium state despite the arbitrariness of the depicted lines (Ma, 2019).

For a long time, medical establishments have used chloral hydrate as a sedative. The chloral hydrate reactant and the hydronium ion and chloride ion that is produced throughout the reaction process include the addition of a proton to the water molecule. The equation for the reaction is $CHCl_3$ + H_2O H_3O+ + Cl-. 58% of the reactant is produced via the hydrate formation process, and the remaining 42% is brought to equilibrium. The "Mick Finns" and "Knockout" tactics are at their disposal for use against rivals. Aldehydes may create hydrates in the presence of alcohol, which is a property shared by water. Ketones can be used to create hemiketals. Alcohol is used in the chemical process instead of water. Contrary to the process of creating esters from carboxylic acids and alcohols, water is not removed from reactive intermediates. Hemicacetal production does not need the elimination of water, in contrast to ester formation. Hemicetal synthesis is the process of combining an alcohol and an aldehyde inside the same molecular structure to produce a cyclic structure that resembles an ester. After experiencing loop formation around the alcohol functional group, a molecule can participate in a reaction with an aldehyde. Establish a circuit with six people in it. Could you possibly provide me with more specific background information or details regarding the ring you are referring to? The phrase "bonding elements" describes chemical substances that have a propensity to combine with other substances to produce more stable compounds. The methyl group on the sixth carbon of the aldehyde is located externally in the ring structure. The hydrogen cation, represented by the symbol H+, is capable of creating a ring. One that is unique or unmatched.

When the ring is closed, the initial carbon atom's orientation of the OH group may take on a cis or trans configuration about the CH_3 group situated outside. When new rings are formed, one of two geometric isomers is randomly generated. It is well known that two different geometric isomers have an impact on the process of ring closure in monosaccharides, including glucose. The six-membered hemiacetal ring of glycosamine displays stability. The condensation process between the aldehyde group of glucose and the hydroxyl group at carbon 5 results in the creation of C. The hydroxyl (OH) group is positioned on the fifth carbon (#5 C) atom as a result of rotating the ring, which facilitates its contact with the carbon atom that bears a partial positive charge. There are two unique OH group structures for the carbon atom at position C1. C could get messages from outside sources, which might be cis or trans. Based on their chemical makeup, carbohydrates are divided into two groups. Cis is not a kind of glucose, in contrast to trans- (Ma, 2019).

The two hemispheres remain in a condition of equilibrium. In an aqueous solution, glucose has aldehyde contents of 36%, 64%, and 0.022%. For the course of the exam, it is critical to have this chemical reaction fresh in your mind. As required, the structure must be shown with appropriate dimensions. Contrary to glucose and other 6-membered rings with single bonds, which are often represented in the chair conformation, which is thought to be the most thermodynamically stable shape, the 6-membered ring of the hemiacetal structure is described in Haworth notation as flat. The beta glucose atomic structure is shown in the diagram. Galactose, a different kind of sugar present in milk, is perhaps produced by a similar chemical mechanism. The fourth carbon position of the glucose molecule has a hydroxyl group, which gives it its distinctive properties. Similar to glucose, chain modification's orientation is important, while it is not necessary.

In the twelfth stage, the carbon-4 hydroxyl group is rotated 180 degrees to restart the glucose synthesis process. Galactose's hydroxyl group is still present in Carbon 4. In contrast to a trans-configuration, the C4 hydroxyl group of a cyclic hemiacetal produced from galactose displays a cis structure concerning C6. A binary option is offered by this arrangement. Make a large amount of galactose synthetically. Nucleic acids like RNA and DNA are formed from ribose. Aldehydes are the precursors of galactose and glucose. The structural configuration of the C-1 hydroxyl group, like that of glucose, determines whether it is in an alpha or beta form. In contrast to alpha ribose, which has the hydroxyl group on carbon #1 orientated in a trans-configuration about carbon, beta ribose has the hydroxyl group on carbon #1 oriented in a cis configuration.

Notwithstanding the 5-membered ring's effect on the shape of the molecules. The noncyclic ribose C ring, which starts the numbering of rings, was first identified as an aldehyde (C=O). Imine and water are the byproducts of the reaction between an amine and an aldehyde. The aforementioned chemical reaction starts similarly to how esters are made. The outcomes are listed below. Formaldehyde has the power to destroy harmful viruses and bacteria found in vaccinations. Forming imines through a reaction between protein amine groups and the aforementioned chemical molecule is how formaldehyde interacts with viral functional domains. The development of antibodies and the elimination of viruses are of the highest significance. Flu, whooping cough, DTaP, hepatitis A/B, and polio vaccinations may include the ingredient formaldehyde. The issue vaccinations probably contain formaldehyde, a substance that might be harmful. A good fixative for histology specimens is formaldehyde because it may crosslink amine groups in proteins. Through its interaction with proteins present in live bacteria and fungi, formaldehyde can preserve corpses.

OSHA controls worker exposure to formaldehyde, a low-level human carcinogen that is frequently present in glues used in wood products. The graphic picture demonstrates the occurrence of an Amadori rearrangement on the carbon atom next to the carbon-nitrogen (C=N) imine functional group when there are an imine moiety and nearby hydroxyl (OH) groups. The resultant chemical has functional groups for ketone and amine that may take part in the complex Maillard process. The Maillard reaction, a chemical process that results in the production of appealing tastes and fragrances in baked products, includes the interaction between amino acids in proteins and reducing sugars in carbohydrates. The distinctive brown hue seen in a variety of foods, including roasted coffee, toast, bread crust, dark brews, and caramel, is a result of the Maillard reaction. The stratum corneum produces colored imine melanoidins when dihydroxyacetone and amine are combined in a self-tanner. Etymologically speaking, the word "melanin" is where the name "melanoidin" comes from. Melanoidins differ structurally, yet they all have similar absorption characteristics. Alcohol, ketone, and erythrulose are examples of potential constituents in DHA-based self-tanning solutions. What physical properties do erythulose and DHA have? The topic of discussion is the chemical molecule known as erythrulose. Protein amine groups may allow glucose to gradually and on its produce imines, according to the phenomena of glycated proteins. For some imines, the Amadori rearrangement may make Maillard reactions more difficult. Glycation happens to proteins. If an organism has glycated proteins or not. Blood glucose levels and the amount of glycated proteins show an inverse connection. Glycated proteins are a side effect of uncontrolled diabetes. Proteins that have undergone glycation are linked to health issues in people with diabetes (Ma, 2019).

CHAPTER

8 Nomenclature of Aromatic Compounds

Contents

8.1. INTRODUCTION

Aromatic compounds are organic compounds with a unique molecular structure characterized by a conjugated π electron system. The term "aromatic" originated from compounds known for their distinct fragrances, but it has evolved to describe a specific chemical property. Aromatic compounds typically consist of one or more cyclic rings, often composed of carbon atoms, with alternating single and double bonds. The presence of this conjugated system imparts exceptional stability and reactivity to aromatic compounds, leading to their significance in various chemical processes (Smith, 2012) (Figure 8.1).

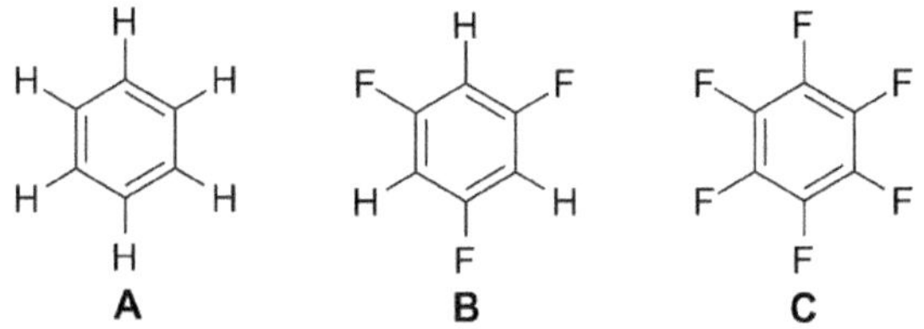

Figure 8.1. Aromatic compounds A-C.

Source: *Estarellas Carolina, Frontera Antonio, Quiñonero David, & Dey Pere, (2010); Scientific Figure on ResearchGate.*

Nomenclature is crucial in chemistry as it provides a standardized system for naming and identifying compounds. The field of organic chemistry deals with an enormous variety of compounds, and having a systematic approach to naming is essential for effective communication and comprehension among chemists. Without a proper nomenclature system, there would be confusion and ambiguity when referring to different compounds, hindering research, collaboration, and the accurate representation of chemical structures.

In the aromatic compounds' case, nomenclature's importance is particularly significant. Aromatic compounds exhibit diverse structures and functional groups, making it necessary to have a well-defined system that accurately represents their composition and arrangement. A systematic nomenclature system enables chemists to communicate specific structural information, aiding in aromatic compounds' synthesis, analysis, and understanding. Furthermore, nomenclature facilitates the organization and classification of aromatic compounds based on their structural features, which is crucial for studying their properties, reactions, and applications in various fields such as pharmaceuticals, materials science, and environmental chemistry.

8.2. HISTORICAL OVERVIEW OF AROMATIC COMPOUNDS

8.2.1. Discovery of Aromaticity

Aromaticity emerged in the 19th century when chemists observed distinct properties in certain compounds. “aromatic” originally referred to compounds with pleasant fragrances but was later associated with specific chemical properties. The discovery of benzene by Michael Faraday in 1825 and its subsequent investigation by several chemists laid the foundation for understanding aromatic compounds. Chemists noticed that benzene exhibited unique stability and reactivity compared to other compounds (Smith, 2012) (Figure 8.2).

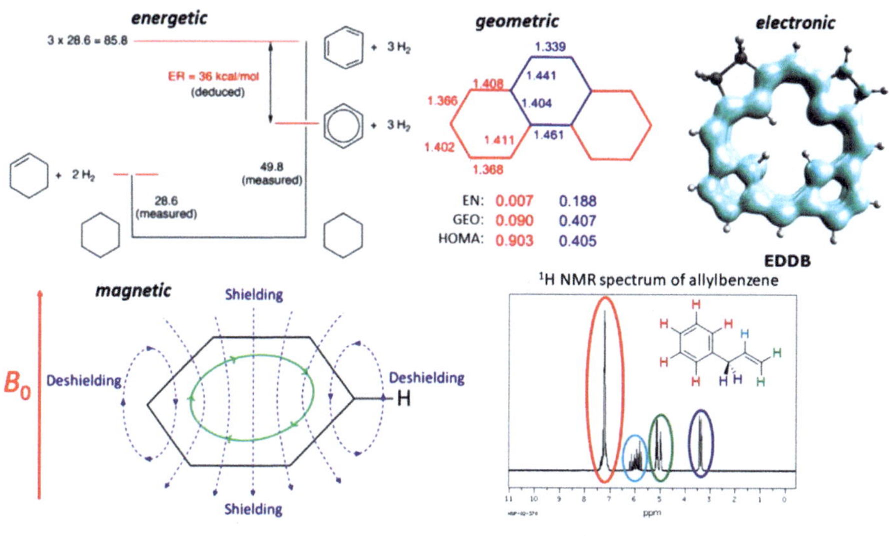

Figure 8.2. Energetic, geometric, electronic, and magnetic properties of molecules used for the assessment of aromaticity.

Source: *Szatylowicz, Wieczorkiewicz, & Krygowski, (2022); MDPI AG.*

Further studies by chemists such as August Wilhelm von Hofmann and Adolf von Baeyer revealed that benzene underwent substitution reactions with remarkable ease, unlike other unsaturated compounds. This behavior was attributed to the stability of the benzene molecule.

The breakthrough in understanding aromaticity came with observing the unique spectroscopic properties of aromatic compounds. In the late 19th century, chemists noticed that aromatic compounds exhibited distinctive absorption spectra in the *ultraviolet* (UV) and infrared (IR) regions, distinguishing them from non-aromatic compounds. These spectroscopic characteristics provided valuable insights into aromatic compounds' electronic structure and bonding.

Sir Robert Robinson and Erich Hückel further advanced the concept of aromaticity in the early 20th century. Hückel's π-electron theory provided a mathematical framework to describe the stability of aromatic compounds based on the number of π-electrons in the cyclic system. This theory laid the foundation for understanding aromaticity beyond benzene and expanded to larger aromatic systems.

The discovery of aromaticity and recognition of its distinct properties were pivotal in shaping our understanding of organic chemistry. These observations set the stage for the subsequent contributions of Kekulé and the development of a comprehensive nomenclature system for aromatic compounds, ensuring accurate representation and effective communication within the field (Guitián, 2013).

8.2.2. Contributions of Friedrich August Kekulé

Friedrich August Kekulé, a German chemist, played a pivotal role in elucidating the structure and nature of aromatic compounds. In 1865, he proposed the Kekulé benzene structure, which depicted a hexagonal ring of carbon atoms with alternating single and double bonds. Kekulé's proposal resolved the long-standing mystery of benzene's structure and provided crucial insights into the concept of aromaticity.

Kekulé's contributions extended beyond benzene. He proposed that aromatic compounds possessed a cyclic structure with a fully conjugated π electron system, which conferred exceptional stability and reactivity. His ideas revolutionized the field of organic chemistry and laid the foundation for understanding the unique properties of aromatic compounds.

Kekulé's work on the structure of benzene was a significant milestone in the field of chemistry. His proposal of alternating single and double bonds in benzene's hexagonal ring explained the compound's experimental observations and provided a structural basis for understanding the stability and reactivity of aromatic compounds (Guitián, 2013).

Kekulé's concept of a fully conjugated π electron system in aromatic compounds was groundbreaking. He recognized that the delocalization of π electrons throughout the cyclic system contributed to aromatic compounds' exceptional stability and unique properties. This insight opened doors to further investigations into aromaticity and paved the way for developing aromatic compound theories, including *Hückel's π-electron theory*. Kekulé's contributions had a profound impact on the field of organic chemistry. His work laid the groundwork for understanding the behavior of aromatic compounds, influencing research in synthesis, structure determination, and reactivity studies.

The Kekulé benzene structure became a fundamental concept in organic chemistry textbooks as a starting point for chemists to explore the diverse world of aromatic compounds (Smith, 2012).

In recognition of his contributions, Kekulé is widely regarded as one of the pioneers of modern organic chemistry, and his work on aromatic compounds remains highly influential.

8.2.3. Development of Nomenclature System for Aromatic Compounds

The development of a nomenclature system for aromatic compounds progressed alongside the understanding of their structure and properties. Common names were used to identify aromatic compounds based on their sources or distinctive characteristics. However, as the field of organic chemistry advanced, there arose a need for a standardized and systematic nomenclature system.

The International Union of Pure and Applied Chemistry (IUPAC) has played a significant role in developing and maintaining a unified nomenclature system for organic compounds, including aromatics.

The IUPAC system provides guidelines and rules for naming aromatic compounds based on their substituents, functional groups, and structural features. The system ensures clarity, precision, and consistency in the naming of aromatic compounds, enabling effective communication and unambiguous representation of their structures (Guitián, 2013).

The historical overview of aromatic compounds encompasses the discovery of aromaticity, the contributions of Friedrich August Kekulé in elucidating their structure and nature, and the development of a systematic nomenclature system. These milestones have paved the way for our understanding and communication of aromatic compounds, enabling advancements in various areas of chemistry and related fields.

8.3. BASICS OF NOMENCLATURE OF AROMATIC COMPOUNDS

8.3.1. Rules and Guidelines for Nomenclature

The nomenclature of aromatic compounds follows the rules and guidelines established by the IUPAC. These rules ensure consistency and clarity in naming aromatic compounds. Some key principles include:

Identify the parent hydrocarbon: The parent hydrocarbon is typically the longest continuous chain of carbon atoms in the molecule, including the aromatic ring. It serves as the basis for naming the compound.

Numbering the carbon atoms: The carbon atoms in the ring are numbered consecutively, starting from any position providing the lowest set of substituents.

Naming substituents: Substituents attached to the aromatic ring are named using specific prefixes or suffixes based on the functional groups present. Common substituents include alkyl groups *(methyl, ethyl), halogens (fluoro, chloro)*, and nitro groups.

Indicating the position of substituents: The locants (numbers) of substituents are included as prefixes before the parent hydrocarbon name. Multiple substituents are listed alphabetically, and if multiple substituents are of the same kind, numerical prefixes indicate the number of substituents (Smith, 2012).There are guidelines for handling special cases in aromatic nomenclature. For example, when multiple substituents exist on the aromatic ring, the locants are separated by commas. If two or more substituents exist at the same carbon atom, the locant is repeated with a comma between them.

Specific suffixes indicate functional groups attached to the aromatic ring. For instance, a carboxylic acid group is denoted by the suffix -carboxylic acid, while an alcohol group is represented by the suffix -ol. When multiple functional groups are present, the appropriate suffixes are used based on the priority order defined by the IUPAC rules. It is important to note that the nomenclature of aromatic compounds also considers stereochemistry, if applicable. The configuration of chiral centers or geometric isomerism is indicated using the R/S or E/Z system. By following these rules and guidelines, chemists can name aromatic compounds accurately and unambiguously, allowing for effective communication and precise structural representation and facilitating research, synthesis, and analysis in aromatic chemistry (Guitián, 2013).

8.3.2. Prefixes and Suffixes in Aromatic Nomenclature

Aromatic nomenclature utilizes specific prefixes and suffixes to indicate the presence of substituents and functional groups. Some common examples include:

- ***Alkyl Groups***: Alkyl substituents are denoted by prefixes such as methyl, ethyl, propyl, etc.
- ***Halogens***: Halogens (fluorine, chlorine, bromine, iodine) are indicated with the prefixes *fluoro-, chloro-, bromo-*, and *iodo-*.
- ***Nitro Group***: The prefix nitro- indicates a nitro group (-NO_2).
- ***Amino Group***: An amino group (-NH_2) is represented by the suffix -amino.
- ***Carbonyl Group***: A carbonyl group (C=O) is denoted by the suffix -one or -aldehyde, depending on its position in the molecule.

  There are additional prefixes and suffixes commonly used in aromatic nomenclature to describe various functional groups and substituents:

- ***Alkoxy Groups***: Alkoxy substituents are indicated by prefixes such as methoxy, ethoxy, propoxy, etc. These groups are derived from alkyl groups attached to an oxygen atom (-O-alkyl).
- ***Aryl Groups***: Aryl substituents are aromatic rings treated as substituents. Examples include phenyl (C_6H_5-) and naphthyl ($C_{10}H_7$-).
- ***Hydroxy Group***: A hydroxyl group (-OH) is indicated by the suffix -ol. For example, phenol is the name for benzene with a hydroxyl group attached (Smith, 2012).
- ***Aldehyde Group***: An aldehyde functional group (CHO) is denoted by the suffix -al or -aldehyde. For instance, benzaldehyde refers to benzene with an aldehyde group attached.
- ***Acetyl Group***: The acetyl group (CH_3CO-) is represented by the prefix acetyl-. For example, acetophenone denotes a compound in which the acetyl group is attached to the phenyl ring.

These prefixes and suffixes, along with the rules for numbering and indicating substituents, provide a comprehensive system for naming aromatic compounds, ensuring accuracy and consistency in their representation.

8.3.3. Common Aromatic Compounds and Their Names

Aromatic compounds encompass a wide range of molecules with distinct names. Some common aromatic compounds and their names include:

- ***Benzene (C_6H_6):*** The simplest aromatic compound, benzene, is the parent hydrocarbon for many other aromatic compounds.
- ***Toluene ($C_6H_5CH_3$)***: A monosubstituted benzene, toluene is named a benzene derivative, with the methyl group indicated by the prefix tolyl-.
- ***Phenol (C_6H_5OH):*** An aromatic compound with a hydroxyl group attached to the benzene ring. The name phenol indicates the presence of the hydroxyl group.
- ***Aniline ($C_6H_5NH_2$):*** An aromatic compound with an amino group ($-NH_2$) attached to the benzene ring. The name aniline signifies the presence of the amino group.

These are just a few common aromatic compounds, and their names adhere to the rules and guidelines of aromatic nomenclature.

By following the established rules and utilizing the appropriate prefixes and suffixes, chemists can accurately name aromatic compounds, allowing for effective communication and a clear representation of their structures and functional groups (Guitián, 2013).

8.4. NAMING MONOSUBSTITUTED AROMATIC COMPOUNDS

8.4.1. Naming Benzene Derivatives

- ***Substituents and their Common Names***: When naming benzene derivatives, it is important to understand the common names given to various substituents. Common substituents include methyl, ethyl, propyl, fluorine, chlorine, bromine, and nitro. These substituents are often referred to by their specific names, such as tolyl for the methyl group ($-CH_3$), ethyl for the ethyl group ($-C_2H_5$), and chloro for the chlorine group (-Cl). Familiarizing oneself with these common names is essential for accurately naming benzene derivatives.
- ***IUPAC Rules for Naming Monosubstituted Benzene Compounds***: The IUPAC rules provide a systematic approach

for naming monosubstituted benzene compounds. The key steps involve identifying the parent hydrocarbon (benzene), assigning the locants (numbers) to the substituents on the benzene ring, and naming the substituents using appropriate prefixes or suffixes. The locants are chosen to provide the lowest set of numbers for the substituents.

- For example, if there is a methyl group at position 3 on the benzene ring, the compound would be named *3-methylbenzene*. If there are multiple substituents, they are listed alphabetically, and if there are multiple substituents of the same kind, numerical prefixes are used (e.g., 1,2-dimethylbenzene).

It is important to consider the impact of substituents on the aromatic ring's reactivity and properties. The position of the substituent can affect the overall stability, electronic distribution, and potential for the reaction of the aromatic compound. Understanding these factors can provide insights into the compound's behavior and interactions with other molecules.

Furthermore, it is essential to note that while common names are widely used, the IUPAC rules provide a standardized and internationally recognized system for naming aromatic compounds. Adhering to the IUPAC rules ensures consistency and clarity in communication among chemists worldwide (Smith, 2012).

By mastering the nomenclature of monosubstituted aromatic compounds, chemists can effectively communicate the structure, composition, and functional groups present in a compound. This knowledge facilitates aromatic chemistry research, analysis, and synthesis, enabling scientists to study these compounds' properties, reactions, and applications with precision and accuracy (Guitián, 2013).

8.4.2. Examples and Practice Problems

It is helpful to work through examples and practice problems to grasp the nomenclature of monosubstituted aromatic compounds further. By encountering various compounds with different substituents, one can become proficient in applying the IUPAC rules and identifying the appropriate names for these compounds.

Practice problems may involve identifying the parent hydrocarbon, assigning locants to the substituents, and correctly naming the compound. These problems allow chemists to hone their skills in nomenclature and gain confidence in naming monosubstituted aromatic compounds.

By understanding the rules and guidelines and practicing with examples, chemists can become proficient in naming monosubstituted aromatic compounds accurately. The ability to correctly name compounds enhances communication within the scientific community and facilitates the understanding and recognition of different compounds in research, education, and industrial applications.

Working through examples and practice problems is an effective way to solidify one's understanding of the nomenclature of monosubstituted aromatic compounds. By encountering a variety of compounds with different substituents, chemists can apply the IUPAC rules and develop proficiency in naming these compounds accurately.

Practice problems often involve identifying the parent hydrocarbon, assigning locants to the substituents, and correctly naming the compound based on the rules discussed earlier. These exercises provide hands-on experience and allow chemists to refine their skills in aromatic nomenclature.

Through consistent practice and exposure to different examples, chemists can confidently and accurately enhance their ability to name monosubstituted aromatic compounds. This proficiency in nomenclature enables effective communication within the scientific community, as researchers can precisely convey the structure and composition of aromatic compounds in their work (Smith, 2012).

Furthermore, a strong grasp of aromatic compound nomenclature facilitates the understanding and recognition of different compounds in various domains, including research, education, and industrial applications. Accurate and standardized naming ensures clarity and consistency, enabling scientists to communicate and collaborate on projects involving aromatic compounds effectively.

8.5. NAMING DISUBSTITUTED AROMATIC COMPOUNDS

8.5.1. Ortho-, Meta-, and Para- Positioning

In disubstituted aromatic compounds, there are three possible arrangements of substituents on the benzene ring: ortho (o), meta (m), and para (p). These terms describe the relative positions of the substituents with respect to each other (Guitián, 2013).

- ***Ortho (o):*** This refers to the substituents occupying adjacent positions on the benzene ring. For example, if there are two substituents at positions 1 and 2, they are considered ortho to each other.
- ***Meta (m):*** Meta refers to the substituents occupying positions 1 and 3, one carbon atom apart on the benzene ring.
- ***Para (p):*** Para describes the substituents occupying positions 1 and 4 directly opposite each other on the benzene ring.

Understanding the ortho, meta, and para positions is crucial for accurately naming disubstituted aromatic compounds, as it allows for the proper arrangement of substituents in the compound's name (Smith, 2012) (Figure 8.3).

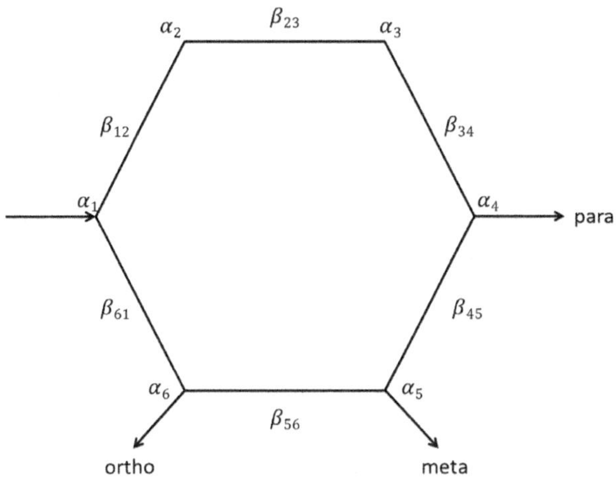

Figure 8.3. Modified benzene molecule, showing para, meta, and ortho connections to leads.

Source: *Qiu Dong, & Sulston Kenneth, (2016); Scientific Figure on ResearchGate.*

8.5.2. IUPAC Rules for Naming Disubstituted Benzene Compounds

The IUPAC rules provide guidelines for naming disubstituted benzene compounds systematically. The key steps involve identifying the parent hydrocarbon (benzene), assigning locants (numbers) to the substituents on the benzene ring, and arranging the substituents in the name based on their positions.

When naming disubstituted benzene compounds, the substituents are listed alphabetically with their locants indicated. The locants are chosen to provide the lowest set of numbers for the substituents, considering both ortho, meta, and para positions.

For example, if there is a methyl group at position 1 and a chlorine group at position 4 on the benzene ring, the compound would be named 1-chloro-4-methylbenzene.

8.5.3. Examples and Practice Problems

To reinforce the understanding of naming disubstituted aromatic compounds, working through examples and practice problems is essential. These exercises involve identifying the positions of the substituents on the benzene ring and correctly arranging and naming the compound according to the IUPAC rules.

Practice problems may include different combinations of substituents and require determining the ortho, meta, or para positions and assigning appropriate locants. By solving these problems, chemists can gain proficiency in naming disubstituted aromatic compounds accurately (Guitián, 2013).

Additionally, practicing with examples enhances the ability to visualize the structure and spatial arrangement of the substituents on the benzene ring. This skill is crucial when encountering more complex disubstituted aromatic compounds.

Through consistent practice and exposure to various examples, chemists can develop confidence and proficiency in naming disubstituted aromatic compounds. Accurate and systematic naming facilitates effective communication and recognition of these compounds in scientific research, education, and industrial applications.

8.6. NAMING POLYSUBSTITUTED AROMATIC COMPOUNDS

8.6.1. Locating Substituents on a Benzene Ring

When dealing with *polysubstituted* aromatic compounds, the challenge lies in determining the positions of multiple substituents on the benzene ring. Locating substituents involves identifying the positions of each substituent and understanding their relationships to one another.

To locate substituents on a benzene ring, it is important to number the carbon atoms systematically. The numbering should provide the lowest set of locants for the substituents. By carefully examining the structure of the compound, chemists can determine the positions of each substituent relative to one another.

Locating substituents on a benzene ring requires a systematic approach to numbering the carbon atoms. The aim is to assign the lowest set of locants to the substituents. To achieve this, chemists often start numbering from a position that allows for the lowest possible numbers for the substituents.

Several factors can influence the decision on where to begin numbering. These include functional groups or substituents that require the lowest possible locants and follow any priority rules for specific substituents. Careful analysis of the compound's structure is crucial in determining the appropriate numbering scheme.

Chemists must consider the relationships between substituents when locating them on the benzene ring. This involves visualizing the spatial orientation and determining the relative positions of the substituents to one another. By understanding these relationships, chemists can accurately identify the locations of multiple substituents and proceed with naming the compound (Smith, 2012).

Properly locating substituents on a benzene ring is essential for correctly applying subsequent nomenclature rules. It ensures the systematic and unambiguous naming of *polysubstituted* aromatic compounds, facilitating effective communication and understanding within the scientific community (Guitián, 2013).

8.6.2. Priority Rules for Numbering Substituents

In *polysubstituted* aromatic compounds, prioritizing rules for numbering the substituents is necessary. These rules ensure consistency and allow for a systematic compound naming approach.

The priority rules typically involve assigning lower locants to substituents based on alphabetical order. If the substituents are of the same kind, numerical prefixes indicate the number of each substituent. By following these priority rules, chemists can accurately determine the numbering of substituents and ensure the correct naming of the compound.

Priority rules are crucial in determining the numbering of substituents in polysubstituted aromatic compounds. These rules establish a hierarchy for assigning locants, ensuring a consistent and systematic approach to naming. The priority rules primarily rely on alphabetical order. When multiple substituents are present, they are listed alphabetically to determine their order of priority. The substituent with the lowest alphabetical name is assigned the lowest locant number. This rule ensures a standardized method for naming compounds and minimizes ambiguity.

In cases where substituents are of the same kind, numerical prefixes indicate the number of each substituent. For example, if two methyl groups are attached to the benzene ring, they would be denoted as "1,2-dimethyl." This numbering system provides clarity and accurately identifies the substituents' positions.

By adhering to priority rules, chemists can establish a consistent naming convention for polysubstituted aromatic compounds. This approach facilitates effective communication, allows for easy comparison of compounds, and aids in interpreting chemical structures in research, education, and various applications.

8.6.3. IUPAC Rules for Naming Polysubstituted Benzene Compounds

The IUPAC rules provide a systematic framework for naming polysubstituted benzene compounds. These rules consider the locations of multiple substituents on the benzene ring and the priority rules for numbering them.

To name polysubstituted benzene compounds, chemists must identify the parent hydrocarbon (benzene) and assign locants to each substituent. The substituents are listed alphabetically, using their locants and appropriate prefixes or suffixes to indicate their positions and functional groups (Guitián, 2013).

For example, if there is a methyl group at position 1, an ethyl group at position 3, and a bromine atom at position 4 on the benzene ring, the compound would be named 1-methyl-3-ethyl-4-bromobenzene.

By adhering to the IUPAC rules, chemists can consistently and accurately name polysubstituted aromatic compounds. This systematic approach ensures effective communication and understanding of the compound's structure and composition in scientific research, education, and industrial applications.

The IUPAC rules provide comprehensive guidelines for naming polysubstituted benzene compounds. These rules consider the positions of multiple substituents on the benzene ring and the priority rules for numbering them (Smith, 2012).

To name polysubstituted benzene compounds, chemists first identify the parent hydrocarbon, which is benzene in this case. Locants, or numbers, are then assigned to each substituent based on the priority rules. The substituents are listed alphabetically, using their locants and appropriate prefixes or suffixes to indicate their positions and functional groups.

For instance, if there is a methyl group at position 1, an ethyl group at position 3, and a bromine atom at position 4 on the benzene ring, the compound would be named 1-methyl-3-ethyl-4-bromobenzene.

By adhering to the IUPAC rules, chemists ensure consistent and accurate naming of polysubstituted aromatic compounds. This systematic approach facilitates effective communication and understanding of the compound's structure and composition across various scientific disciplines and applications.

8.6.4. Examples and Practice Problems

To solidify the understanding of naming polysubstituted aromatic compounds, working through examples and practice problems is invaluable. These examples provide real-world scenarios and allow chemists to apply the IUPAC rules for naming.

Practice problems may involve identifying the parent hydrocarbon, assigning locants to substituents, and correctly naming the compound. By actively engaging with examples and practice problems, chemists can enhance their proficiency in naming polysubstituted aromatic compounds and build confidence in their skills.

Working through examples and practice problems allows chemists to encounter a variety of compounds with different substituents, reinforcing their understanding of the rules and applying them in different contexts. This practice also helps chemists develop a systematic approach to tackle complex compounds by breaking them down into individual substituents and assigning locants based on priority rules.

Examples and practice problems challenge chemists to think critically and apply their knowledge of nomenclature principles. By encountering different scenarios, chemists become familiar with the nuances of naming polysubstituted aromatic compounds, such as dealing with multiple substituents of the same kind or prioritizing functional groups (Smith, 2012).

Moreover, practice problems provide an opportunity to identify common mistakes and misconceptions, allowing chemists to refine their understanding and improve accuracy in compound naming. Through repetition and exposure to various compounds, chemists gain proficiency and fluency in the nomenclature of polysubstituted aromatic compounds.

8.7. NAMING FUSED AROMATIC COMPOUNDS

8.7.1. Nomenclature of Bicyclic Aromatic Compounds

Fused aromatic compounds are characterized by two or more aromatic rings that share common carbon atoms. Naming these compounds requires a systematic approach to accurately and concisely represent their structures.

Regarding bicyclic aromatic compounds, the nomenclature is based on naming individual aromatic rings and indicating their fusion. The fused rings are named individually, and their fusion positions are indicated using appropriate prefixes (Guitián, 2013) (Figure 8.4).

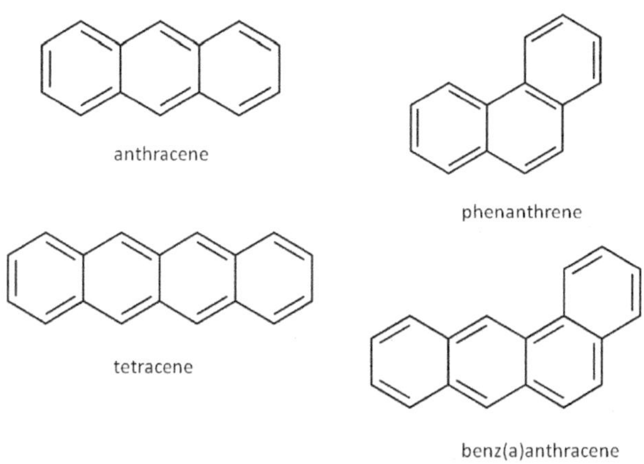

Figure 8.4. Compounds containing fused aromatic rings or polycyclic aromatic hydrocarbons (PAH); Their properties vary depending on the number of the fused aromatic rings and their arrangement.

Source: *Gheisari Ali, (2017); Scientific Figure on ResearchGate.*

The first step in naming bicyclic aromatic compounds is identifying the parent hydrocarbon, typically the larger or more complex ring. The substituents attached to the parent ring are named using the rules and guidelines discussed earlier, considering their positions on the individual rings. The fusion of the rings is indicated using specific prefixes that describe the position and nature of the fusion. Common prefixes include "ortho," "meta," and "para" to indicate fusion at adjacent, 1-position, or 1,2-positions, respectively. Additionally, numerical locants indicate the substituents' positions on each ring.

8.7.2. Examples and Practice Problems

To reinforce the understanding of naming fused aromatic compounds, working through examples and practice problems is crucial. Examples may include compounds with different combinations of fused aromatic rings and various substituents (Smith, 2012).

By encountering different examples, chemists can practice applying the nomenclature rules specific to fused aromatic compounds. This includes identifying the parent rings, determining the fusion positions, assigning locants to substituents, and constructing the appropriate compound name.

Practice problems offer an opportunity to refine naming skills and solidify knowledge. These problems may involve naming compounds with multiple fused rings, multiple substituents, or complex fusion patterns. By actively engaging with such problems, chemists enhance their ability to accurately and consistently name fused aromatic compounds.

Through examples and practice problems, chemists can gain confidence and proficiency in the nomenclature of bicyclic and fused aromatic compounds. This enables effective communication and understanding of the structural characteristics of these compounds in research, education, and various applications in the field of chemistry (Guitián, 2013).

8.8. NAMING HETEROCYCLIC AROMATIC COMPOUNDS

8.8.1. Nomenclature of Heterocycles with One Heteroatom

Heterocyclic aromatic compounds in the aromatic ring contains at least one heteroatom, an atom other than carbon. Naming these compounds follows specific rules to represent their structures and identify the heteroatom accurately.

When naming heterocycles with one heteroatom, it is named as a substituent using its common name or an appropriate suffix. The parent hydrocarbon is usually benzene; a prefix denotes the heteroatom before the parent hydrocarbon name.The heteroatom prefixes commonly used include "oxo-" for oxygen, "*thio*-" for sulfur, "*aza*-" for nitrogen, and so on. For example, a heterocycle with a nitrogen atom in the ring is named by prefixing "*aza-*" to the parent hydrocarbon name, such as "*azobenzene.*"

The systematic naming of heterocyclic compounds with one heteroatom allows chemists to describe their structures and compositions accurately. This naming convention ensures clear communication and facilitates recognizing and understanding different heterocyclic compounds in various scientific disciplines, including pharmaceuticals, agrochemicals, and materials science. By following the nomenclature guidelines, chemists can effectively convey the specific arrangements of atoms in these important classes of compounds (Smith, 2012).

Naming heterocycles with one heteroatom provides a foundation for understanding and naming more complex heterocyclic systems with multiple heteroatoms. It establishes a consistent approach that allows chemists to navigate the nomenclature of increasingly diverse and intricate heterocyclic aromatic compounds.

8.8.2. Nomenclature of Heterocycles with Multiple Heteroatoms

Each heteroatom is named as a substituent using the appropriate prefix or suffix in heterocycles with multiple heteroatoms. The numbering of the heteroatoms follows the same principles as in other aromatic compounds, providing the lowest set of locants for the substituents.

The heteroatoms are named using prefixes that indicate their identity, such as "oxy-" for oxygen, "*thio-*" for sulfur, and "*aza-*" for nitrogen. For example, a heterocycle with oxygen and nitrogen atoms in the ring can be named "*oxyazabenzene.*"

In cases with multiple heteroatoms of the same type, numerical prefixes indicate the number of each heteroatom. For instance, a heterocycle with two oxygen atoms in the ring can be named "*dioxybenzene.*"

The systematic naming of heterocycles with multiple heteroatoms allows for precisely representing their structures and compositions. It ensures that each heteroatom is accurately identified and located within the aromatic ring, facilitating effective communication and understanding among chemists (Guitián, 2013).

Understanding the nomenclature of heterocycles with multiple heteroatoms is crucial in various fields, including medicinal chemistry, where many drugs are based on heterocyclic structures. Accurate and consistent naming practices enable researchers to discuss and analyze the structures of complex heterocyclic compounds, aiding in the development of novel therapeutic agents and materials with desired properties.

8.8.3. Examples and Practice Problems

To reinforce the understanding of naming heterocyclic aromatic compounds, examples, and practice problems are valuable tools. Examples present real-world heterocycles with different heteroatoms, allowing chemists to apply the nomenclature rules. Practice problems may involve identifying the parent hydrocarbon, assigning locants to the heteroatoms, and correctly naming the compound. By actively engaging with examples and practice problems, chemists can enhance their proficiency in naming heterocyclic aromatic compounds and develop confidence in their skills (Smith, 2012). Working through examples and practice problems in heterocyclic nomenclature enables chemists to become familiar with the naming conventions and improve their ability to accurately communicate the structures and compositions of heterocyclic aromatic compounds. This proficiency is crucial in various fields, including medicinal chemistry, materials science, and organic synthesis. By encountering different examples and practicing with various heterocyclic structures, chemists can broaden their knowledge and become adept at handling complex naming scenarios. These exercises also enhance problem-solving skills and critical thinking as chemists learn to analyze the structures, identify the heteroatoms, and apply the appropriate nomenclature rules. Ultimately, effectively naming heterocyclic aromatic compounds allows chemists to communicate their research findings, collaborate with peers, and contribute to advancements in diverse areas, such as drug discovery, agrochemicals, and developing novel materials with specific properties.

8.9. NAMING AROMATIC COMPOUNDS WITH FUNCTIONAL GROUPS

8.9.1. Introduction to Functional Groups in Aromatic Compounds

Functional groups are specific arrangements of atoms within a molecule that confer characteristic chemical properties and reactivity. When functional groups are present in aromatic compounds, naming these compounds requires considering both the aromatic ring and the functional group (Guitián, 2013). The functional group is treated as a substituent on the parent hydrocarbon in aromatic compounds with functional groups. Identifying the functional group and its position relative to the aromatic ring is important. Common functional groups found in aromatic compounds include hydroxyl groups (-OH), amino groups (-NH_2), carbonyl groups (-C=O), and carboxyl groups (-COOH) (Smith, 2012).

Functional groups in aromatic compounds can significantly affect their chemical behavior, solubility, and biological activity. Therefore, accurately naming these compounds is crucial for effective communication within the scientific community and for understanding their properties and potential applications.

By incorporating information about the specific functional group and its position on the aromatic ring, chemists can convey precise details about the structure and reactivity of aromatic compounds. This facilitates collaboration, research dissemination, and advances in various scientific fields, including organic chemistry, medicinal chemistry, and materials science.

8.9.2. IUPAC Rules for Naming Aromatic Compounds with Functional Groups

The IUPAC rules provide guidelines for naming aromatic compounds with functional groups, ensuring systematic and consistent nomenclature. These rules consider the position of the functional group on the aromatic ring and any additional substituents present.

To name aromatic compounds with functional groups, the parent hydrocarbon, usually benzene, is identified. The functional group is then named as a substituent, using appropriate prefixes or suffixes according to its nature. The locant, or position, of the functional group on the aromatic ring, is indicated by a numerical prefix.

In some cases, when multiple functional groups are present, priority rules may apply to determine substituents' numbering and ensure the compound is named correctly. By following the IUPAC rules, chemists can accurately represent the structure of aromatic compounds with functional groups and facilitate effective communication and understanding.

By adhering to the IUPAC rules, chemists can effectively represent the structure of aromatic compounds with functional groups, enabling clear communication and facilitating a better understanding of their properties, reactivity, and potential applications. The systematic and consistent nomenclature provided by these rules is essential for scientific research, education, and the advancement of various fields, including organic chemistry, pharmaceutical sciences, and chemical engineering (Guitián, 2013).

8.9.3. Examples and Practice Problems

To reinforce the knowledge of naming aromatic compounds with functional groups, examples, and practice problems are invaluable resources. Examples showcase real-world aromatic compounds with different functional groups, allowing chemists to apply the nomenclature rules and gain proficiency in naming (Smith, 2012).

Practice problems may involve identifying the parent hydrocarbon, assigning locants to the functional groups and any additional substituents, and correctly naming the compound. Through active engagement with examples and practice problems, chemists can enhance their skills in naming aromatic compounds with functional groups and develop confidence in their abilities.

Working through examples and practice problems enables chemists to encounter various aromatic compounds with various functional groups, consolidating their understanding of the naming conventions and guidelines. This practice also enhances their problem-solving skills and ability to interpret and communicate the structural features of aromatic compounds with functional groups.

Proficiency in naming aromatic compounds with functional groups is essential in many areas of chemistry, including organic synthesis, pharmaceutical research, and developing specialty chemicals. Accurate and concise naming facilitates effective communication, collaboration, and documentation of research findings, ensuring the advancement of knowledge and innovation in diverse scientific fields.

By actively practicing and applying the IUPAC rules, chemists can confidently navigate the complex naming of aromatic compounds with functional groups. They become adept at recognizing and assigning appropriate names to these compounds, enabling efficient communication and understanding among researchers, educators, and industry professionals. Using examples and practice problems is a stepping stone to proficiency, allowing chemists to strengthen their skills and expand their knowledge of aromatic compounds with functional groups.

8.10. SPECIAL CASES IN AROMATIC NOMENCLATURE

8.10.1. Naming Aromatic Compounds with Common Names

Aromatic compounds are sometimes referred to by their traditional or trivial names rather than following the strict IUPAC nomenclature. These common names have historical significance and are widely recognized and used in various fields, including chemistry, biochemistry, and pharmacology.

Common names often stem from the compound's historical importance, origin, or significant applications. For example, benzene (C_6H_6) is a well-known aromatic compound with a ring of six carbon atoms, and its common name has become the accepted term for this compound. Similarly, compounds like toluene, naphthalene, and anthracene have commonly used names that have become deeply rooted in the scientific literature (Guitián, 2013).

While common names lack IUPAC nomenclature's systematic and standardized approach, they continue to be used for convenience and familiarity. Chemists and researchers must know the common and corresponding IUPAC names to ensure effective communication and understanding in scientific discussions (Smith, 2012).

In addition to benzene, toluene, naphthalene, and anthracene, many other aromatic compounds have common names that are widely recognized. For example, phenol is the common name for benzene with a hydroxyl (-OH) functional group attached, while aniline refers to benzene with an amino (-NH_2) group. These common names are deeply ingrained in scientific literature and are often used in academic and industrial settings.

Using common names for aromatic compounds allows for easier communication and accessibility, particularly in fields where these compounds play a significant role. Common names are often used in everyday conversations, practical applications, and the pharmaceutical industry, where drug names are frequently based on common aromatic compound names.

However, it is important to note that while common names are widely recognized, they do not provide the same precision and clarity as the IUPAC systematic nomenclature. The IUPAC rules ensure a consistent and unambiguous naming system, especially when dealing with complex structures or when international collaboration is involved.

8.10.2. Naming Aromatic Compounds with Multiple Rings

Aromatic compounds can possess multiple aromatic rings, which introduces additional complexity to their nomenclature. When naming aromatic compounds with multiple rings, each ring is considered individually, and the prefixes ortho-, meta-, and para- are used to indicate the relative positions of the rings (Guitián, 2013).

The prefixes ortho- (o-), meta- (m-), and para- (p-) describe the spatial arrangement of the rings. Ortho- signifies that the rings are adjacent, meta- indicates a 1,3-substitution pattern, and para- represents a 1,4-substitution pattern.

The naming of aromatic compounds with multiple rings also involves assigning locants to the substituents on each ring and considering the overall arrangement and connectivity of the rings. It is essential to accurately identify and locate the substituents on the individual rings to ensure the precise representation of the compound's structure.

In the nomenclature of aromatic compounds with multiple rings, the prefixes ortho-, meta-, and para- play a crucial role in indicating the relative positions of the rings. These prefixes provide valuable information about the spatial arrangement and connectivity of the aromatic rings within the compound.

When two rings are adjacent, they are referred to as ortho-, denoted by the prefix o-. This means the substituents or functional groups are attached to carbon atoms directly next to each other on the two adjacent rings. Similarly, meta- (m-) signifies a 1,3-substitution pattern, where the substituents or functional groups are attached to carbon atoms separated by one carbon atom on each ring. On the other hand, para- (p-) represents a 1,4-substitution pattern, with the substituents or functional groups attached to carbon atoms that are four positions apart on the rings (Smith, 2012). Assigning locants, or numerical labels, to the substituents on each ring is crucial in accurately representing the compound's structure. The locants help distinguish the specific positions of the substituents relative to the aromatic rings, enabling a precise and unambiguous description of the compound.

When dealing with aromatic compounds with multiple rings, carefully considering the positional relationships, locants, and the ortho-, meta-, and para- prefixes ensures a systematic and standardized approach to their nomenclature. This facilitates effective communication and comprehension of the compound's structure and properties within the scientific community (Guitián, 2013).

8.10.3. Naming Aromatic Compounds with Substituents on Substituents

In complex aromatic compounds, it is possible to have substituents attached to other substituents rather than directly to the parent hydrocarbon. When naming aromatic compounds with substituents on substituents, the IUPAC rules extend to multiple levels, considering the substituents' nested nature.

A hierarchical approach is followed to name aromatic compounds with substituents on substituents. The primary substituent is named first, followed by the secondary substituent, and so on until all the substituents are accounted for. Each substituent level is named using appropriate prefixes or suffixes, and locants are assigned to indicate their positions.

Naming aromatic compounds with substituents on substituents requires careful attention to detail and adherence to the hierarchical order of substituents. Accurately representing the structure and connectivity of these compounds ensures effective communication and understanding among chemists.

When naming aromatic compounds with substituents on substituents, a hierarchical approach is employed to ensure a systematic and concise nomenclature. The process involves naming each substituent level in a nested manner, starting from the primary substituent and proceeding to the secondary substituent, tertiary substituent, and so on, until all the substituents are accounted for.

At each substituent level, the appropriate prefixes or suffixes are used to denote the nature of the substituents. Locants are assigned to indicate the positions of the substituents within the compound, considering their nested arrangement. This hierarchical naming approach allows for a comprehensive representation of the compound's structure and the precise location of each substituent. Naming aromatic compounds with substituents on substituents demands attention to detail and adherence to the hierarchical order of the substituents. By following these guidelines, chemists can effectively communicate the complex structure of these compounds and facilitate a clear understanding among researchers in various scientific disciplines.

Accurate and systematic nomenclature is essential in organic synthesis, drug discovery, and materials chemistry, where precise communication of compound structures is crucial. Properly naming aromatic compounds with substituents on substituents ensures clarity, consistency, and accuracy in scientific discussions and publications (Smith, 2012).

8.10.4. Examples and Practice Problems

Solidifying the understanding of naming aromatic compounds with special cases, examples, and practice problems provide valuable opportunities for application and reinforcement. Examples illustrate various scenarios, such as compounds with common names, multiple rings, and substituents on substituents, allowing chemists to practice the corresponding nomenclature.

Practice problems may involve identifying the appropriate nomenclature for specific compounds, considering their unique characteristics, and conforming to the relevant rules and guidelines. By actively engaging with examples and practicing problems, chemists enhance their proficiency in naming aromatic compounds with special cases and develop confidence in their skills.

Working through examples and practice problems enables chemists to encounter various aromatic compounds with diverse structures and naming challenges. This practice reinforces the principles and concepts learned in the nomenclature of aromatic compounds with special cases. Chemists become familiar with the specific considerations and nuances involved in naming compounds with common names, multiple rings, and substituents on substituents (Guitián, 2013).

Examples and practice problems allow chemists to apply their knowledge and problem-solving skills to real-world scenarios. They learn to identify the appropriate nomenclature for different compounds, considering their unique characteristics and conforming to the relevant rules and guidelines. This practical experience strengthens their understanding and mastery of naming aromatic compounds with special cases.

Through active engagement with examples and practice problems, chemists understand aromatic compound nomenclature principles more deeply. They become adept at recognizing patterns, identifying key features, and making accurate and concise nomenclature choices.

Proficiency in naming aromatic compounds with special cases is vital for effective communication and collaboration within the scientific community. The ability to accurately represent compound structures and convey essential information about their composition and functional groups ensures clarity and facilitates further research and advancements in various fields, including organic chemistry, pharmaceuticals, and materials science (Smith, 2012). In conclusion, the nomenclature of aromatic compounds plays a crucial role in accurately representing their structures, identifying functional groups, and facilitating effective communication among chemists. The systematic approach IUPAC rules provide ensures consistency and clarity in naming these compounds.

Throughout this discussion, we have explored various aspects of aromatic compound nomenclature. We started with the fundamentals of naming aromatic compounds based on the number and arrangement of carbon atoms in the ring. We then delved into naming fused aromatic compounds, bicyclic aromatic compounds, heterocyclic aromatic compounds, and aromatic compounds with functional groups.

We examined the principles, rules, and guidelines for assigning appropriate names to these compounds in each case. We discussed the importance of identifying parent hydrocarbons, assigning locants to substituents and heteroatoms, and using prefixes and suffixes to indicate positions and the nature of fusion or functional groups (Guitián, 2013).

Furthermore, we explored special cases such as naming aromatic compounds with common names, multiple rings, and substituents on substituents. These cases require additional considerations and specific approaches to represent the compound structures and accurately communicate effectively.

Examples and practice problems have been highlighted as valuable tools for reinforcing understanding, applying the rules, and developing proficiency in aromatic compound nomenclature (Smith, 2012).

CHAPTER

9 How to Draw Organic Chemical Structures from an IUPAC Name

Contents

9.1. INTRODUCTION

The ability to deduce the structures of organic chemicals from their corresponding IUPAC names is an important skill in organic chemistry. The International Union of Pure and Applied Chemistry (IUPAC) has established a systematic naming system to ensure consistent and clear identification of organic compounds. Chemists who are familiar with the conventions and rules of IUPAC nomenclature can accurately translate the complicated names of organic compounds into molecular structures.

The steps involved in drawing organic chemical structures from IUPAC names are decoding the name to identify the parent chain, functional groups, and substituents, and then representing these elements in a visual representation. This skill is required to understand and explain the composition, reactions, and properties of organic compounds (Pletnev, 2013).

The IUPAC will establish a uniform naming scheme for organic compounds. The IUPAC nomenclature system provides a methodical and clear way of communicating the structure and composition of organic molecules. Understanding the IUPAC name is critical for both chemists and students because it allows for an accurate representation of organic compounds and clear communication. This discussion will go over the complexities of understanding the IUPAC name, including its elements, regulations, and conventions.

9.1.1. The Components of the IUPAC Name

The IUPAC name of an organic compound is made up of several elements that describe its structure and functional groups. The main components are the parent chain, functional groups, and substituents. Let's take a closer look at each of these components.

1. ***The Parental Chain:*** The parent chain is the longest continuously arranged carbon chain in the compound. It establishes the organic molecule's base name and serves as its backbone. The functional groups are prioritized, with the highest priority functional group chosen as the parent chain. The name's prefix (such as meth-, eth-, prop-, but-, etc.) is also determined by the number of carbon atoms in the parent chain (Gomollón-Bel, 2019).
2. ***Organizations in Action:*** Functional groups are atomic clusters that give molecules distinct chemical properties and reactivity. They

are critical when determining how to name organic compounds. Typical functional groups include alcohols, aldehydes, ketones, carboxylic acids, amines, and esters. The IUPAC name contains a suffix or a prefix to indicate the functional group, depending on where it is in the parent chain.

3. ***Substituents:*** Substituents are groups or atoms attached to the parent chain that is not part of the primary carbon skeleton. They are named independently of the parent chain and are identified by prefixes in the IUPAC name. The position of the substituents is indicated using locants, which are numbers that represent the carbon atoms to which the substituents are connected (Pletnev, 2013).

9.1.2. Customs and Regulations

To ensure consistency and clarity in the naming of organic compounds, IUPAC nomenclature follows a set of guidelines and conventions. These rules must be understood for the IUPAC name to be correctly interpreted.

1. ***Numbering the Parent Chain*:** The parent chain is numbered from the beginning to the end, with the lowest number assigned to the substituents. This is done to achieve the lowest locant values for substituents and functional groups. If there are multiple substituents, the numbering is chosen so that the overall number of locants is as low as possible.
2. ***Locants*:** Locants are used to indicate the location of functional groups and substituents on the parent chain. They appear immediately following the prefix or suffix denoting the functional group or substituent, respectively. Locants are separated by commas when there are multiple substituents.
3. ***Priority Order*:** To determine their priority, functional groups are ranked in order of seniority. The compound's base name is determined by the highest priority group. When a compound contains both an alcohol and an aldehyde group, the aldehyde group takes precedence, and the compound is called an aldehyde.

Hyphens and commas are used to separate numbers and words to denote locants or different parts of a name. Commas are used to separate multiple locators or substituents. To ensure clarity and avoid ambiguity in the IUPAC name, hyphens and commas should be used correctly (Gomollón-Bel, 2019).

9.1.3. Exceptions and Special Situations

Although the IUPAC nomenclature system is comprehensive, some exceptions and special cases necessitate extra caution.

1. ***Tautomers***: Tautomers are isomers that interconvert while in equilibrium by the migration of a hydrogen atom. Their connectivity differs, but their molecular makeup is the same. Both tautomeric forms are mentioned in the name in these cases to indicate the compound's dynamic nature.
2. ***Ambiguous Naming***: The name of some compounds can be interpreted in a variety of ways depending on where functional groups or substituents are located. In these circumstances, proper clarification is required to avoid ambiguity in the IUPAC name (Pletnev, 2013).

Knowing the IUPAC name is essential for accurately representing and communicating the structure and composition of organic compounds. By understanding the elements, guidelines, and conventions of the IUPAC nomenclature system, chemistry professors and students can decipher complicated organic names and gain valuable insights into the structure and reactivity of organic molecules. The IUPAC name serves as a common language in the field of organic chemistry, facilitating effective communication as well as further study and advancement in this fascinating field.

9.2. DETERMINING THE PARENT CHAIN

The IUPAC guidelines state that determining the parent chain is a critical step in naming organic compounds. The parent chain is critical in establishing the overall structure and nomenclature because it serves as the foundation for identifying and categorizing the compound. In this discussion, we will look at the method for identifying the parent chain in organic compounds, as well as the variables that influence this choice.

Organic compounds are made up of carbon atoms that have been linked to various functional groups and substituents to form chains or rings. The parent chain, the longest continuous chain of carbon atoms in the compound, serves as the molecule's structural core. It serves as the foundation for naming and outlining the compound's structure. The parent chain is chosen under IUPAC guidelines. The first step in determining the parent chain is to locate the main carbon chain within the compound. This chain can be of various lengths and can be straight or branched. Because this is the parent chain, it is critical to determine which chain has the most carbon atoms. Carbon atoms in the parent chain are typically joined together by single bonds to form an alkane backbone (Gomollón-Bel, 2019).

Several recommendations should be followed to select the correct parent chain. First, the parent chain should include the functional group with the highest priority. For example, if the compound contains both an alcohol (-OH) and an alkene (=C=) functional group, the parent chain should be the one that contains the alkene because it has a higher priority under IUPAC rules.

In addition, the parent chain should strive to have the longest carbon chain possible while incorporating as many substituents as possible. Substituents are atoms or groups of atoms that are attached to the parent chain. Prefixes are used to identify specific functional groups or substituents when they are present. By selecting a parent chain that contains the majority of the substituents, the compound's name is made shorter and more accurately reflects its structural characteristics.

When multiple chains within the compound are the same length, other factors are used to identify the parent chain. According to IUPAC rules, chains with more substituents and functional groups are preferred over chains with fewer of them. The concept of "priority" is important when naming organic compounds because it promotes consistency and uniformity in the nomenclature system.

It is important to remember that each functional group and substituent has its own set of criteria for determining the parent chain. In cyclic compounds, for example, the parent chain may be a ring rather than a linear carbon chain. The number of carbon atoms in the ring determines the parent chain in this case, and the substituents are named accordingly (Pletnev, 2013).

Furthermore, compounds can have multiple functional groups with equal priority. In these cases, the parent chain is chosen based on the functional group that appears first in the name, according to the alphabetical order of the prefixes used to designate these groups.

To demonstrate the procedure for identifying the parent chain, consider the compound with the molecular formula $C_6H_{12}O$, which contains both an alkene and an alcohol group. To properly name this compound, we must select the correct parent chain.In this case, both a straight-chain hexane and a chain with a double bond, which would result in a pentene, are potential parent chains. Because the alkene group has a higher priority than the alcohol group, the pentene will be the parent chain rather than the hexane. By using the pentene as the parent chain, we can accurately represent the structure and give the functional group in the compound's name priority.

To summarize, naming organic compounds using the IUPAC nomenclature system necessitates careful consideration of the parent chain. It entails selecting the carbon chain with the longest continuous length, the most substituents, and the highest priority functional group. It is possible to correctly identify the parent chain by following specific guidelines and considering factors such as priority, substituents, and functional groups. Understanding the parent chain concept is essential for naming and categorizing organic compounds, as well as communicating effectively within the organic chemistry community (Gomollón-Bel, 2019).

9.3. IDENTIFYING FUNCTIONAL GROUPS

9.3.1. Alkanes, Alkenes, Aromatic Rings, and Alkynes

The hydrocarbon functional groups are highly non-polar and extremely weak acids. In contrast to molecules with more polar functional groups, hydrocarbons' only intermolecular interactions are London dispersion forces, and their boiling points are typically quite low. Alkanes are hydrocarbons with only one bond. Alkane substituents are alkyl groups, which are alkanes without a C-H bond, such as methyl, ethyl, or propyl.

Alkanes include methane, ethane, propane, butane, and octane. Alkanes have a high covalent C-H bond and a very low polarity. Water does not dissolve them. Alkyl carbons are sp^3 hybridized and have tetrahedral geometry around the carbon.

In organic chemistry, alkyl groups, denoted by the symbol R-, are frequently referred to as "spectator" functional groups. Except for free-radical substitution and, of course, combustion, alkanes do not undergo many different reactions. They are frequently used as the nucleus of organic molecules.

Alkenes are hydrocarbons with one or more double bonds formed between carbon atoms. Common examples include ethene, propene, and butene. Alkene substituents are known as alkenyl groups, and $-CH=CH_2$ is also known as vinyl.

Alkenyl carbons are sp^2 hybridized and have a trigonal planar geometry. Alkynes have a triple carbon and carbon bond. Ethyne (acetylene) is the most basic alkyne; acetylenes are another name for alkynes. Alkyne substituents are referred to as alkynyl groups. Terminal alkynes have a C-H bond. Internal alkynes have C-C bonds at both ends. Linear geometry sp hybridization is performed on alkynyl carbons. Benzene rings are composed of six-membered rings with three double bonds. Benzene rings are abundant in nature due to their unusual stability and a property known as aromaticity (which has nothing to do with smell). It could be represented as a hexagon with a circle on top. Toluene, also known as benzene, is responsible for the distinct odor of model airplane glue. C_6H_5- substitutes for benzophenone are phenyl groups. Benzene contains trigonal planar sp^2 hybridized carbons (Pletnev, 2013).

9.3.2. Alcohols, Ethers, Amines, and Thiols Are Examples of Alkyl Halides

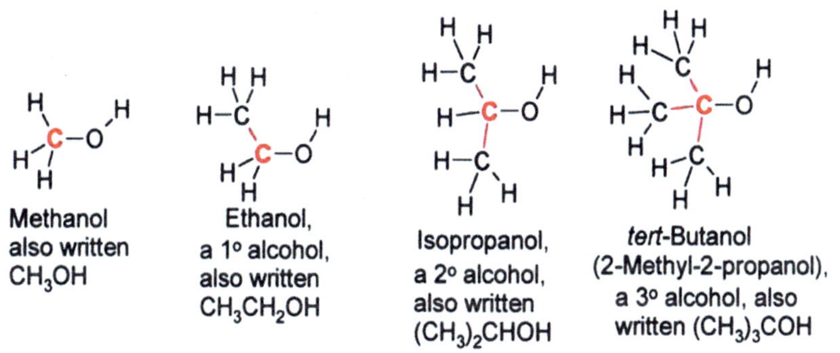

Figure 9.1. Chemical structure of popular alcohols.

Source: *Jarrell John, (2008); Scientific Figure on ResearchGate.*

These functional groups are noticeably more polar due to the greater difference in electronegativities between the bonding atoms. Because of the interactions between the dipoles, these functional groups have stronger intermolecular forces. Alcohols and amines have higher boiling points due to their ability to form hydrogen bonds (Figure 9.1).

Carbon is bonded to the hydroxyl group in alcohols with the prefix R-OH. Common examples include methanol, isopropanol, and methanol. The O-H bond is highly polarized and participates in hydrogen bonding. Furthermore, hydroxyl groups improve water solubility. Alcohols, in addition to being weak acids, can act as Lewis bases. Carboxylic acids, a distinct functional group, are thought to be hydroxyl groups bonded to C=O.

Ethers R-O-R is made up of two carbon bonds that surround oxygen atoms. Dioxane, tetrahydrofuran, and diethyl ether are some of the most commonly used ethers in laboratories. Because ethers cannot act as hydrogen-bond donors, their boiling points are lower than those of comparable molecular-weight alcohols but higher than those of hydrocarbons due to stronger dipole-dipole interactions (Gomollón-Bel, 2019).

R-F, R-Cl, R-Br, and R-I are functional groups for alkyl halides, where R is an alkyl group. Chloroform, bromobutane, and methyl bromide are examples of alkyl halides. Dipole-dipole interactions have higher boiling points than alkanes. If R is an alkene, they are alkenyl halides. Required functional groups for substitution and elimination reactions. Amino acids contain the functional groups -NH_2, -NHR, or NR_2, and R is typically a hydrocarbon. Morphine, codeine, and cocaine are the three most well-known amine-containing molecules. Another name for amino substituents is amino groups. Amines with N-H bonds can form hydrogen bonds, raising their boiling points and increasing their water solubility.A single pair of nitrogen atoms can act as a base. Thiols (mercaptans) R-SH are the sulfur-containing cousins of alcohols. Because the sulfur atom is not nearly as electronegative as oxygen, the S-H bond is much less polarized than the oxygen bond. Thiols are weak acids with higher potency than alcohols. Ethanethiol, which is added to natural gas to give it that distinct aroma, has the strongest stench associated with thiols (Pletnev, 2013).

9.3.3. Esters, Ketones, Aldehydes, and Carboxylic Acids Are All Types of Esters

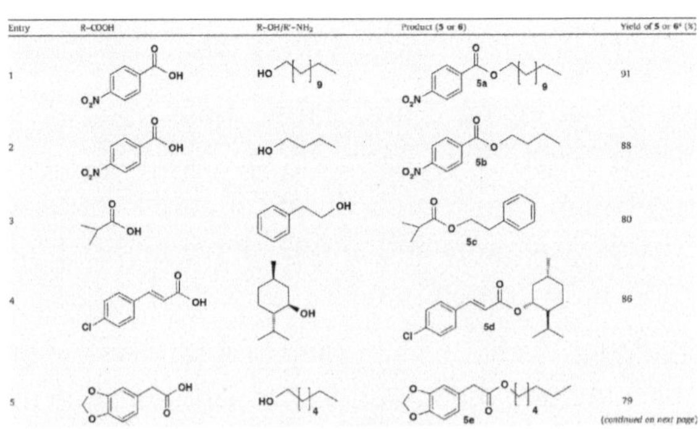

Figure 9.2. Structural formulas of various esters and amides.

Source: *Mcnulty James, Krishnamoorthy Venkatesan, & Robertson Al, (2008); Scientific Figure on ResearchGate.*

The C=O group is also known as the carbonyl group. Carbon is partially positively charged, and the C=O bond is strongly polarized towards oxygen. Carbonyls can be found in aldehydes, ketones, esters, and carboxylic acids. Carbon and C-H are bonded by the compound C=O in aldehydes RCHO (Figure 9.2).

Benzaldehyde, acetaldehyde, and formaldehyde are a few examples. Although they form polar covalent bonds, they cannot donate hydrogen. In the ketone RC(O)R, C=O is linked to two carbons.

To remove nail polish, acetone (2-propanone) is used. In carboxylic acids (RCOOH), a carbonyl is linked to the -OH group. In terms of functional groups, they differ from alcohol (Gomollón-Bel, 2019).

Formic acid and acetic acid (vinegar) are the two most basic carboxylic acids. Other short-chain acids, such as butanoic and pentanoic acids, are notorious for smelling like a locker room. Carboxylic acids have higher boiling points due to the hydroxyl group's involvement in hydrogen bonding. Unlike strong acids such as HCl and H_2SO_4, carboxylic acids, despite their name, typically exhibit partial dissociation in water.

Unlike carboxylic acids, esters RCOOR have an O-C bond instead of an O-H bond. Esters can be identified by their sweet odors. Have polar bonds but do not participate in hydrogen bonds. Acid halides, nitriles, anhydrides, and amides are examples of organic compounds.

Carbonyls are present in a wide range of important functional groups. Because they can be produced by substituting different groups for OH, all of these functional groups are thought to be carboxylic acid derivations. (Although nitriles do not appear to be related to carboxylic acids at first glance, dehydration can convert them to amides.)

Amides are made up of an amino group and a carbonyl carbon. Peptides are amino acid groups that are joined together by the formation of an amide. Hydrogen bonding between amides with N-H bonds is possible (Pletnev, 2013).

In acid halides, F, Cl, Br, or I are substituted for -OH. In anhydrides, oxygen is surrounded by two carbonyls. Not to be confused with esters. The name "anhydrides" comes from the fact that they can be formed by converting two equivalents of a carboxylic acid while also losing water. Nitriles do not appear to be carboxylic acid derivatives at first glance, but they can be made by dehydrating amides. Acetonitrile is a common solvent. "Nitrile gloves" are made of nitrile rubber, a butadiene-acrylonitrile copolymer, and a butadiene-acrylonitrile copolymer. The cyanide designation is sometimes used to describe the -CN substituent. HCN, or hydrogen cyanide, is a hazardous gas. The cyanide ion CN^- is frequently encountered in introductory courses (on paper, not in the lab!).

Nitro, Imine, Azide, and Other Epoxides, Thioethers, and Other. There is no overarching theme here; only a few more noteworthy functional groups to be aware of. Although epoxides are technically ethers, they deserve their classification because they participate in several reactions that ethers do not. Thioethers (sulfides) are the sulfur equivalent of ethers. Dimethyl sulfide is the most frequently mentioned example.

Electrons are strongly attracted to nitro groups. Nitromethane, a solvent, is the most basic nitroalkane. Imines are the nitrogen-containing counterparts of aldehydes and ketones. Azides appear on occasion. The "A" in the anti-HIV medication AZT stands for azido.

9.3.4. The Concept of Functional Groups

Functional Groups (FGs) are collections of linked atoms that control the properties and reactivity of a parent molecule. They are important in organic chemistry, medicinal chemistry, toxicity assessment, spectroscopy, and, last but not least, chemical nomenclature. The study of common FGs constitutes a significant portion of the basic organic chemistry curriculum. A large number of scientific papers and books have been written about the properties and reactivity of various FGs. A well-known example is the classic book series "Chemistry of functional groups," which has over 100 volumes and describes various classes of organic molecules. However, the study of functional groups from the standpoint of cheminformatics receives surprisingly little attention. The majority of theoretical studies use FGs as the foundation for chemical ontologies, with FGs acting as "keys" for categorizing molecules in a hierarchical fashion. Examples of this type of publication include the work of Bobach et al. (2012) who described a rule-based definition of chemical classes to classify compounds, and the ClassyFire software developed in Wishart's group, which allows chemists to perform large-scale automated chemical classification based on a structure-based chemical taxonomy with over 4800 categories (Gomollón-Bel, 2019).

Various substructure features are frequently used in cheminformatics in conjunction with machine learning to create models that predict the biological activity or characteristics of molecules. This method extracts groups of atoms from molecules using a predefined algorithm to generate substructure descriptors. Among these descriptors are topological torsions, linear or atom-centered fragments, and pharmacophoric triplets. Although these fragment descriptors are very useful, they do not describe functional groups. Fragments are generated for all parts of a molecule without regard for their potential chemical function, and they are typically highly overlapping.

Haider's Checkmol (2012) program was one of the first pieces of software capable of recognizing FGs. It recognized approximately 200 FGs. The program recently released an expanded version with 583 manually curated SMARTS-encoded functionalities. This list also includes numerous heterocyclic rings and common structural patterns (such as a 5-membered aromatic ring with 1 heteroatom). These substructure features are used in the development of QSAR models for the prediction of toxicity and other molecular physicochemical properties. The well-known ZINC database and related web-based software suite store approximately 500 so-called chemical patterns, which speed up substructure searches and allow estimation of molecule reactivity. Other substructures are included in the patterns, as well as PAINS filters that recognize repeat offenders obstructing biochemical screens. Based on their experience with internal screening campaigns, scientists at Eli Lilly developed another popular set of substructures that are used to recognize potentially reactive or promiscuous molecules. ToxPrint chemotypes are a set of generic chemical functionalities defined recently by the ToxPrint program that describe molecule substructure, reaction features, and atom and bond properties. The tool's primary goal is to use these features in toxicity modeling (Pletnev, 2013).

There is no computer program that we are aware of that can recognize functional groups (FGs) in a molecule without relying on a manually curated set of substructure features. The algorithm in this article is an attempt to address this scientific problem.

9.3.5. Identifying and Eliminating Functional Groups

The majority of FGs contain heteroatoms. As a result, our strategy involves processing heteroatoms and their surroundings while also incorporating additional functionalities, such as multiple carbon-carbon bonds.

The algorithm is described as follows:

1. Mark all heteroatoms in a molecule, including halogens.
2. Also, emphasize the following carbon atoms:
    - Atoms connected by a non-aromatic double or triple bond to any heteroatom

- Atoms in nonaromatic double or triple carbon-atom bonds.
- Acetal carbons are sp^3 carbons that have been linked to two or more oxygen, nitrogen, or sulfur atoms. These O, N, or S atoms must have only one bond.
- Every atom found in the rings of oxirane, aziridine, and thiirane (due to their high reactivity, these rings are commonly regarded as functional groups).

3. All marked connected atoms should be combined into a single FG.
4. Extract FGs with unmarked carbon atoms that are not part of the FG but are instead part of the FG's environment.

The aforementioned algorithm only cycles through non-aromatic atoms. Aromatic heteroatoms are gathered as individual atoms rather than as a larger system. They are only extended to a larger FG when an aliphatic functionality is connected (for example, an acyl group connected to a pyrrole nitrogen). Heteroatoms found in heterocycles are typically regarded as a component of the entire heterocyclic ring rather than as "classical" FGs in and of themselves. This treatment is justified by the wide variety of heterocyclic systems. In our previous study, for example, we counted nearly 600,000 different heterocycles, each with one to three fused 5- and 6-membered rings (Gomollón-Bel, 2019).

Following the above-mentioned marking of all FG-related atoms, the identified FGs are extracted along with their surroundings, i.e., there are connected carbon atoms when the type of carbon (aliphatic or aromatic) is also preserved.

We do not claim that this algorithm is the last word on FGs. An FG is most likely understood differently by medicinal chemists. The definition of activated sp^3 carbons, in particular, may spark debate. We did not consider other similar systems, such as alpha-substituted carbonyls or carbons connected to S=O or similar bonds, in the current algorithm, instead limiting our definition to classical acetal, thioacetal, or aminal centers (i.e., sp^3 carbons with at least two oxygens, sulfurs, or nitrogens as neighbors). During the program development phase, many of these options were tested, and the "strict" definition produced the best results. By extending FGs to alpha-substituted carbonyls (i.e., heteroatom or halogen in the alpha position to carbonyl) and related systems, the number of FGs identified more than tripled, resulting in the generation of numerous large and uncommon FGs. We used this strict definition of acetal carbons because we were comparing different molecular datasets rather than estimating reactivity. To assess the potential reactivity of molecules, several substructure filters, such as the previously mentioned PAINS or Eli Lilly rules, are available.

9.3.6. Increasing the Number of Functional Groups

FGs, particularly those with multiple connection points, can take on a variety of shapes depending on their surroundings. The attachment points can be linked to aliphatic or aromatic carbons in a variety of ways, or they can be left unsubstituted (the valences are filled with hydrogens). A simple amide group with three connection points can generate 18 of these variations (here, two nitrogen connections are considered symmetrical) (Pletnev, 2013).

However, going into such detail is not always necessary. When analyzing the frequency statistics of FGs in chemical databases, one is usually more interested in the proportion of molecules that have functionalities such as urea or sulfonamide rather than the environmental specifics. As a result, combining FGs based on the critical "central" moiety would be preferable. However, one must exercise caution in this situation. The distinction between alcohols and phenols, or amines and anilines, is critical in some unusual situations, particularly for smaller FGs. The following generalization scheme was created to account for these various scenarios:

1. Carbon atom environments are removed, except for carbonyl substituents, which are retained (to distinguish between aldehydes and ketones).
2. The "R atoms" (this atom may represent hydrogen or carbon) fill all open valences on heteroatoms except:
   - Hydrogens bound to -OH groups
   - Simple amines and thiols (FGs with a single central N or S atom) do not have their hydrogens replaced, allowing secondary and tertiary amines, thiols, and sulfides to be distinguished.
3. All remaining environment carbons (on heteroatoms and carbonyls) are replaced by "R atoms" except for environments on single-atomic N or O FGs with one connected carbon; this allows for the differentiation of amines and anilines, as well as alcohols and phenols (Gomollón-Bel, 2019).

This method strikes a good balance between generalization and the retention of relevant chemical details. The following section contains examples of generalized FGs generated by this procedure.

9.3.7. Analysis and Findings

The FGs for the bioactive portion of the ChEMBL database (2017), which contains approximately 483,000 molecules with activity below 10 um on any ChEMBL target, were identified and collected using the above-mentioned algorithm, which was then implemented using the Novartis internal cheminformatics system written in Java. Before discarding organometallic structures, all molecules were standardized by removing counterions and neutralizing atomic charges.

Amide is the most common FG, accounting for 41.8% of all molecules. The most common substituents are ester groups (37.8%), tertiary amines (25.4%), fluoro (19.0%), and chloro (18.5%). In total, our process discovered 3080 FGs. Their distribution is a typical power law "long tail" distribution (similar to the distribution of other substructure features such as substituents or linkers), with a few very common groups and a large number of infrequent groups. 33 FGs are found in more than 1% of molecules, 88 are found in more than 0.1% of molecules, and 1218 (39.5%) are singletons (found in only one molecule).

To validate this novel approach, a comparison with the most advanced program currently available for FG identification—check mol—was performed. Checkmol analyzes molecules and selects FGs from a set of 204 functionalities. Checkmol defines FGs in a hierarchical fashion, which differs slightly from our philosophical approach. We used Checkmol's most intricate substructure as a comparison. In the ChEMBL dataset, the check mol algorithm discovered 165 FGs. The results show a strong relationship between the two approaches. The differences are usually less than a tenth of one percent. Checkmol typically identifies slightly fewer FGs for simple aromatic substituents than our method (for example, aryl chloride 18.13 vs. 18.17% or phenol 9.52 vs. 9.58%), which is likely due to the two systems' slightly different definitions of aromaticity. The most significant differences between the two approaches can be seen in the simple functionalities that are frequently included in the more complex FGs. The most notable example is alkene, which has a checkmol value of 8.63% but only 4.46% when measured using our method. The reason for this is that the alkene is frequently a component of more common, more complex FGs (such as acrylamide, acryloyl, and so on) that checkmol does not recognize but that our algorithm does. Another example is a ketone, where the acyl functionality is frequently a component of more complex FGs, and the acylhydrazone alone accounts for 1.17% of the difference (7.82% by checkmol, 5.31% by our approach), where the acyl functionality is a component of the FG. Our algorithm discovered that the most common FGs are acylhydrazone, which occurs in 1.17% of ChEMBL structures but is not on the checkmol list, acrylamide, acryloyl, acylsulfonamide, and acrylester (Pletnev, 2013).

Finally, there is a very good agreement between the two methods. This demonstrates how systems based on a carefully selected set of substructures, such as checkmol, can provide useful insights into the FG composition of large molecular datasets. One of our algorithm's advantages over competing systems is its ability to handle more complex FGs, which are frequently found in specialized collections of molecules. This straightforward analysis reveals activated alkenes, which frequently serve as covalent binders and Michael acceptors. One advantage of the algorithmic approach over predefined sets of substructures is the previously mentioned detailed classification of areas, which may be important for analyzing properties such as solubility. Another advantage is the ease with which the algorithm can be fine-tuned to meet the needs of a specific analysis (for example, library diversity or QSAR studies) (Gomollón-Bel, 2019).

A brand-new algorithm for locating all functional groups in organic molecules is presented. Because the algorithm is relatively simple and complete instructions with examples are provided, incorporation into any cheminformatics toolkit should be relatively simple. The author is willing to help in any such endeavor. The new method allows for the analysis of FGs in massive chemical databases, which was previously impossible. Many of these studies are currently underway, intending to identify functionalities common to various classes of natural products as well as differences in FG distribution between bioactive and "average" molecules.

9.4. DRAWING ORGANIC STRUCTURES

Because a molecular formula does not always accurately represent a single compound, structural formulas for organic compounds must be developed. Isomers are different compounds with the same molecular formula, and the prevalence of organic isomers reflects carbon's extraordinary versatility in forming stable bonds with other elements as well as itself. When the atoms that make up the molecules of various isomers are bonded together in fundamentally different ways, we refer to such compounds as constitutional isomers. The structural formulas for each of the seven constitutional isomers of $C_4H_{10}O$ are shown in the table below. All known and hypothetical $C_4H_{10}O$ compounds are represented by these formulas, which share a common structural element. There are no double or triple bonds in any of these structures, nor are there any rings (Figure 9.3).

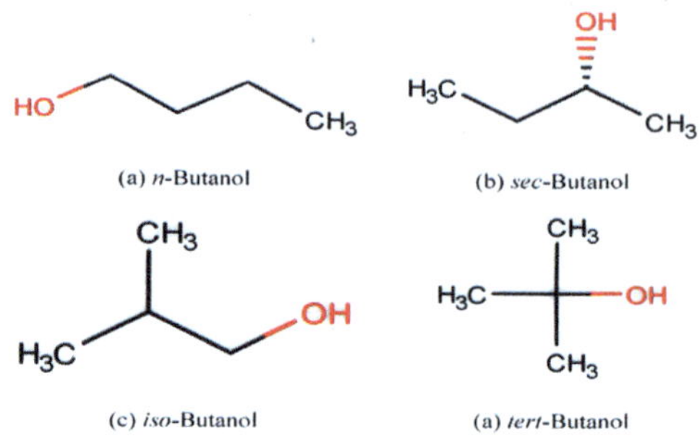

Figure 9.3. Isomers of butanol and their chemical structures.

Source: *Sahoo DrRajesh, Das Aradhana, Gaur Mahendra, Dey Suchanda, Sahoo Saubhagini, & Subudhi Enketeswara, (2019). Scientific Figure on ResearchGate.*

Structure formulas can be simplified without losing any of their information. Condensed structural formulae do not include the bonds that connect each carbon, but each distinct structural unit (group) is still written with subscript numbers designating multiple substituents, including hydrogens.

All references to the elements carbon and hydrogen are omitted in formulas with line bonds (skeletal). The correct number of hydrogens is calculated using carbon tetravalency, where each straight lines segment represents a bond and the ends and intersections of the lines represent carbon atoms. Non-bonding valence shell electrons are excluded from these formulas (Pletnev, 2013).

To become proficient at visualizing a three-dimensional structure from a two-dimensional formula, it takes practice and, in most cases, the use of molecular models. As previously mentioned, the beginning student should purchase one of the many types of model kits available to students and professional chemists.

9.4.1. Kekulé Formulas

The atoms of a molecule are shown in the order in which they are bonded in a structural formula. It also depicts the various types of covalent bonds between atoms, such as single, double, and triple bonds. Covalent bonds are represented by lines. The number of dashes in a bond determines whether it is a single, double, or triple covalent bond. Structural formulas are useful because empirical and molecular formulas frequently fail to adequately represent the properties and structure of the compound (Gomollón-Bel, 2019).

9.4.2. Condensed Formulae

Condensed structural formulas, which save space and are easier and faster to write out, display the atom positions on a single line, similar to a structural formula. Condensed structural formulas are also useful for demonstrating how a group of atoms in a compound are linked to a single atom. When this happens, the group of atoms is enclosed in parenthesis to show that they are grouped.

For example, ethanol's condensed structural formula is CH_3CH_2OH (C_2H_6O in its molecular form).

9.4.3. Line-bond (Skeletal) Formulae

Because organic compounds can be complex at times, line-angle formulas are used to write carbon and hydrogen atoms more effectively. Every point where one line intersects another contains a carbon atom. Carbon's four bonds are then assumed to be completed by hydrogen atoms.

All of the atoms that are linked to the carbon atoms are listed. Line angle formulas, like structural formulas, have advantages and disadvantages in that they help reveal the structure and order of the atoms in a compound (Figure 9.4).

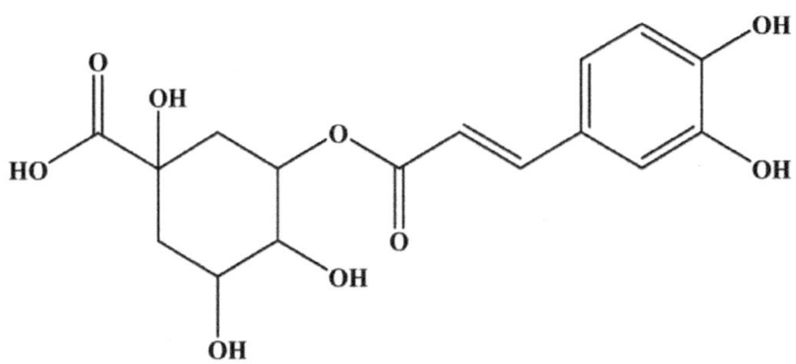

Figure 9.4. Line-bond formulae of chlorogenic acid.

Source: *Shahidi Fereidoon, Varatharajan Vamadevan, Oh Won, & Peng Han, (2019); Scientific Figure on ResearchGate.*

9.5. HANDLING STEREOCHEMISTRY: A COMPREHENSIVE GUIDE TO UNDERSTANDING AND REPRESENTING MOLECULAR CONFIGURATIONS

Stereochemistry is the study of the three-dimensional arrangement of atoms in molecules and how this arrangement affects the characteristics and behavior of compounds. It is critical for understanding the structure, reactivity, and biological function of organic molecules. This manual will examine the principles of stereochemistry with a focus on the various aspects of molecular configurations, such as chirality, stereoisomerism, and the various ways of representing and naming stereoisomers.

9.5.1. Chirality and Stereoisomers

The chemistry of life is heavily reliant on organic compounds, which are molecules built around a chain of carbon atoms (also known as the carbon backbone). The energy carried by these molecules, primarily in the form of potential energy between atomic molecules, is what gives them significance. Understanding the concept of an isomer, a molecule with the same atomic composition as another but different structural arrangements, is critical because changes in atomic placement can have a significant impact on the potential force. This section will concentrate on the chirality of stereoisomers, a type of isomer (Pletnev, 2013).

Chirality is a fundamental concept in stereochemistry. A molecule is chiral if it cannot be superimposed on its mirror image. The chiral center is a central carbon atom that is linked to four different substituents. This asymmetry results in the formation of two distinct mirror images, or enantiomers. Enantiomers are physically identical except for their optical activity (physical interaction with plane-polarized light).

The concepts of stereoisomerism and chirality are heavily emphasized in modern organic chemistry because they aid in understanding the theoretical and physical underpinnings of the formation and structure of numerous organic molecules, which is the primary source of the energy contained in these vital chemicals. In contrast to more well-known constitutional isomerism, which develops isotopic compounds simply by different atomic connectivity, stereoisomerism typically maintains equal atomic connections and orders of building blocks, as well as the same numbers of atoms and types of elements.

So, what makes stereoisomers so distinct? To respond to this question, the learner must be able to think and visualize in three dimensions in addition to two dimensions. Because stereoisomers differ from other isomers in the spatial arrangement of their atoms, they are isomers.

9.5.2. Organization of Space

To comprehend stereoisomerism and chirality, one must first grasp the concept of spatial arrangement. The spatial arrangement of atoms refers to the positioning of various atomic particles and molecules around an organic compound, specifically its carbon chain. In this sense, even a small amount of movement of an atom in any three-dimensional direction will change the spatial arrangement of an organic molecule. This results in an enormous number of potential molecules, each with its unique arrangement of atoms in three dimensions (Gomollón-Bel, 2019).

9.5.3. Stereoisomers

As previously stated, stereoisomers contain various isomer types, each of which has distinct characteristics that further distinguish them as distinct chemical entities with distinct properties. The below mirror-image stereoisomers are classified as enantiomers, which will be discussed in depth.

9.5.4. Enantiomers

This is the fundamental mirror-image, non-superimposable type of stereoisomer introduced at the beginning of the article. The molecules are chiral and achiral are compared. (a) An asterisk denotes the stereocenter of the chiral molecule bromochlorofluoromethane. The original structure cannot be produced by rotating its mirror image. To superimpose the mirror images, bonds must be broken and reformed. When rotated, dichlorofluoromethane and its mirror image are superimposable (Pletnev, 2013). Even if the left molecule is flipped over to the right, the atomic spatial arrangement will not be equal. In molecules, this is referred to as "handedness" and is analogous to the relationship between the left and right hands. Because it may appear counter-intuitive at first, this article suggests that readers try the "hand" example. Place both palms up next to each other. Switch from one side to the other. The palm of one hand and the back of the other should be visible. They are not the same and cannot be combined. As one of the fundamental and defining concepts of stereoisomerism, the concept of chirality enters the picture.

9.5.5. Chirality

When a molecule is chiral, it means that its mirror image (which it must have) is not identical to itself. The definition of chirality is "mirror-image, non-superimposable molecules." Whether a molecule is chiral or achiral is determined by several overlapping factors. The achiral molecule contains two atoms of the same element, which is a distinguishing feature.

In general, no molecule is achiral if the atoms surrounding it differ. The idea of chirality, which is used to explain one type of stereoisomerism, is simple but important. The chemical properties of the chiral molecule differ from those of its mirror image, which is why chirality is important in modern organic chemistry (Gomollón-Bel, 2019).

9.5.6. Several Chiral Center Compounds

The following section focuses on molecules with multiple stereocenters. Our starting point will be D-erythrose, a typical four-carbon sugar.

The stereocenters C2 and C3, both of which have the R configuration, demonstrate the chiral nature of D-erythrose. You should also build a model to convince yourself that, regardless of conformation, finding a plane of symmetry through the molecule is impossible. Is there a D-erythrose enantiomer? Of course, if it's a chiral molecule, it has to. L-erythrose is the name of the enantiomer of erythrose that is its mirror image (again, use models to convince yourself that these erythrose mirror images are not superimposable).

Diastereomers are molecules that are stereoisomers by definition (same molecular formula, same connectivity, but different atom arrangement in space) but are not enantiomers. In practice, this means that a pair of diastereomers has at least one opposite chiral center, but not necessarily all of them. Two diastereomers are not, by definition, mirror images of one another.

In the enantiomer of D-threose known as L-threose, the R configuration is found at C2, and the S configuration is found at C3. L-threose is a diastereomer of both erythrose enantiomers (Pletnev, 2013).

A structure with n stereocenters will usually have 2n different stereoisomers. We will not consider the stereochemistry of double bonds at this time; that will come later. Keep in mind that many sugar molecules can exist in both open-chain and cyclic forms. Consider the glucose molecule in its open-chain form as an example. The two enantiomers of glucose are

D-glucose and L-glucose. The D-enantiomer is the typical sugar that our bodies use as fuel. Because D-glucose has n = 4 stereocenters, there are 2n = 24 = 16 potential stereoisomers.

9.5.7. Selecting Chirality

To determine whether a molecule is chiral, several conditions must be met. First, the molecule must have a chiral center, which is typically a carbon atom bound to four different substituents. Second, due to the placement of these substituents, there should be no internal plane of symmetry in the molecule. If both requirements are met, the molecule is considered chiral.

9.5.8. R/S Notation

In the Cahn-Ingold-Prelog (CIP) system, the R/S notation is used to represent the absolute configuration of a chiral molecule. In the R/S notation, substituents are prioritized according to their atomic number, with higher atomic numbers taking precedence. By orienting the molecule so that the lowest priority substituent faces away from the viewer, the remaining substituents are ranked from highest to lowest priority. If the priority order is clockwise, the configuration is referred to as R (from the Latin word rectus, which means right). If the priority order is counterclockwise, the configuration is known as S (from the Latin word sinister, which means "left") (Gomollón-Bel, 2019).

9.5.9. Fisher Indication

The three-dimensional arrangements of chiral molecules are shown using a two-dimensional representation known as the Fischer projection. In a Fischer projection, bonds that project outward from the plane are represented by horizontal lines, whereas bonds that project inward from the plane are represented by vertical lines. The highest priority goes to the topmost substituent, and the lowest priority goes to the bottommost substituent. By arranging the substituents in this manner, the chirality of the molecule can be determined.

9.5.10. Enantiomers

Enantiomers are mirror images of one another and have opposite configurations at each chiral center. They have the same boiling, melting, and solubility points, but they behave differently when exposed to plane-polarized light. Optically active enantiomers rotate the plane of polarized light in opposite directions (Pletnev, 2013).

9.5.11. Diastereomers

Stereomers are stereoisomers that do not mirror images of one another. Diastereomers, unlike enantiomers, have different configurations at some but not all chiral centers. They have distinct physical properties and are frequently isolated as distinct compounds. Diastereomers exhibit a variety of biological activities and reactivities.

9.5.12. Isomerism (Trans/Cis)

Cis/trans isomerism, also known as geometric isomerism, occurs when compounds have constrained rotation around a double bond or a ring. Trans isomers differ from cis isomers in that their substituents are on the opposite side of the double bond or ring. Cis/trans isomerism is common in compounds with multiple double bonds or that are cyclic.

9.5.13. Substances that Cycle

Cyclical compounds complicate stereochemistry. They may exhibit geometric isomerism due to the limited rotation around the ring. Furthermore, chiral centers may exist if the cyclic structure contains a carbon atom linked to four different substituents. It is critical to understand the spatial arrangement of the substituents to accurately represent the configurations of cyclic compounds (Gomollón-Bel, 2019).

9.5.14. Representation Methodologies

A variety of techniques are used to represent and visualize stereochemistry in organic molecules. In addition to Fischer projections, the following methods are frequently used:

1. ***Sawhorse Projection*:** Sawhorse projections provide a three-dimensional view of a molecule by displaying the spatial relationship between the substituents and the chiral center. Bonds that extend outward are represented by solid lines, while bonds that extend inward are represented by dashed lines. It is depicted in perspective.
2. ***Newman Projection*:** The Newman projection shows how the substituents of a molecule are arranged conformationally. The

bond between two carbon atoms is shown directly downward, with the atoms and substituents represented as circles and lines, respectively.

3. ***3D Computer Models*:** Technological advances have enabled the creation of three-dimensional computer models that accurately represent the stereochemistry of organic compounds. These models allow for a more immersive and interactive exploration of molecular configurations.

The fundamental concept of stereochemistry in organic chemistry governs the properties and behavior of molecules. Accurate molecular configuration understanding and representation are required for analysis, communication, and predicting reactivity. Understanding the concepts of chirality, stereoisomerism, and the various representational techniques can help chemists better understand the complex world of stereochemistry. With practice and exposure to a variety of examples, one can become proficient in handling stereochemistry, improving their skills in organic synthesis, drug design, and other areas of chemical research (Pletnev, 2013).

CHAPTER

10 Abbreviations, Limitations of the IUPAC System, and Trivial Names

Contents

10.1. INTRODUCTION

Abbreviations, the limits of the IUPAC system, and trivial names are all important parts of the field of chemistry that help scientists talk to each other, name chemicals, and figure out how complex chemical combinations work. This introduction is meant to explain these ideas and why they are important in the area of chemistry. Abbreviations are useful tools for making complicated terms easier to understand and for communicating more effectively in scientific and technical areas. In the field of chemistry, where chemical names are often long and complicated, nicknames help researchers and professionals get their points across quickly and clearly. They make it possible to show complicated terms, formulas, and chemical structures in a more compact way. This saves room in publications and makes it easier for people to share information. In chemistry, some common symbols are pH (potential of hydrogen), deoxyribonucleic acid (DNA), ribonucleic acid (RNA), and adenosine triphosphate (ATP). These shorthand terms are known and understood all over the world, which makes them important for effective communication in the field (Ehrenhauser, 2015) (Figure 10.1).

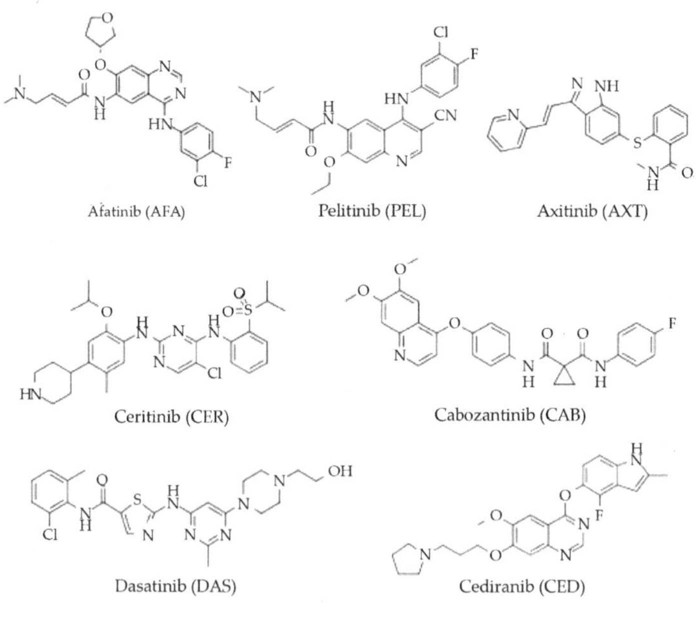

Figure 10.1. Chemical structures of the investigated tyrosine kinase inhibitors (TKIs) with their abbreviations.

Source: *Darwish, Darwish, Ali, & Almutairi, (2023); MDPI AG.*

Even though abbreviations make things easier, it is important to know that the International Union of Pure and Applied Chemistry (IUPAC) system, which is in charge of standardizing chemistry nomenclature, has some flaws. The IUPAC system is a set of rules for naming chemical substances based on how they are made and what they look like. But the method is hard to understand and has some problems. One of the hardest things to do is come up with names for complex organic compounds that can have complicated structures and more than one purpose. When this happens, it can be hard and take a long time to come up with a name, so other ways or trivial names may be used instead (Johnson, 2010) (Figure 10.2).

Abbreviation	Full Name
V_2C MXene	two-dimensional vanadium carbide
V_2AlC	vanadium aluminum carbide
V_2O_3	vanadium trioxide
V_2O_5	vanadium pentoxide
V_8C_7	vanadium carbide
NaV_6O_{15}	sodium vanadium oxide
LiF	lithium fluoride
NaF	sodium fluoride
KF	potassium fluoride
HCl	hydrochloric acid
HF	hydrofluoric acid
NaCl	sodium chloride

Figure 10.2. Sample definition of a chemical abbreviation.

Source: *Wu Meng, Wang Bingxin, Hu Qianku, Wang Libo, & Zhou Aiguo, (2018); Scientific Figure on ResearchGate.*

Trivial names are popular or non-systematic names for compounds that are based on history, geography, or descriptions. Most of the time, these names come from real needs or old customs. They may not follow the strict rules of systematic nomenclature. The use of trivial names goes back hundreds of years, and you can find them in many areas of science. For example, the names "aspirin" (acetylsalicylic acid), "caffeine" (1,3,7-trimethylxanthine), and "quinine" (2-(3-methoxy-4-quinolinyl) ethylamine) are all well-known and accepted names for trivial chemicals. Some trivial names are easier to remember and easier to remember than systematic names, which may be why they are still used in some situations (Ehrenhauser, 2015).

There are some good things about trivial names, but there are also some bad things. When the same chemical is called by different meaningless names in different places or languages, it can cause misunderstanding and miscommunication. This lack of consistency can make it hard for scientists around the world to work together and share their information. Also, trivial names may not give clear information about the chemical structure or traits of a compound. This makes it hard for researchers to figure out how different compounds relate to each other or to predict how they will act. In order to deal with these problems and encourage international standardization, groups like IUPAC try to come up with a method for naming chemical compounds. The goal of these efforts to standardize is to make sure that chemical structures and properties are shown clearly, consistently, and correctly in all languages and areas. By using consistent name rules, the scientific community can improve communication, make things clearer, and make it easier to work together on chemical research and development (Johnson, 2010).

Nicknames, the limits of the IUPAC system, and trivial names are all important parts of chemistry that affect how scientists talk to each other, how they name chemicals, and how they understand complex compounds. Even though abbreviations make it easier to communicate, the limitations of the IUPAC system and the use of meaningless names are problems that need to be fixed for chemical nomenclature to be uniform and standardized. For scientific communication to be clear and accurate and to encourage international cooperation in the field of chemistry, efforts must be made to set up systematic naming standards.

10.2. EXPLORING THE PURPOSE AND SIGNIFICANCE OF ABBREVIATIONS IN SCIENTIFIC AND TECHNICAL FIELDS

Abbreviations are very important in scientific and technical fields because they make complicated terms easier to understand and make it easier for people to talk to each other. Abbreviations are used a lot in chemistry, biology, physics, and engineering, among other fields, because it's important to get across a lot of information quickly. This introductory talk will try to figure out what abbreviations are for and what they mean, focusing on how they help make complicated words easier to understand and improve communication in scientific and technical groups.

One of the main goals of abbreviations is to shorten long words or phrases without changing their meaning. In scientific writing and technical documents, reading and understanding a lot of terms can be confusing and take time. Researchers and experts can explain complicated ideas and concepts more clearly and in less space by using abbreviations. This helps readers understand the information quickly and saves valuable space. In the area of chemistry, for example, abbreviations like DNA, RNA, ATP, and pH (potential of hydrogen) are commonly used and understood, making it easier to talk about basic molecular parts and ideas (Ehrenhauser, 2015).

In the scientific and professional worlds, abbreviations are also a way to use standard language. They let researchers and workers from different places and backgrounds talk to each other well, even if they speak different languages or use different terms. For example, using standard abbreviations at global conferences or in joint research projects makes things clearer and gets rid of any confusion that could come from using different terms. This standardized method makes it easier for people to talk to each other, share ideas, and work together across borders, which leads to improvements in scientific knowledge and new ideas.

Abbreviations are important for more than just making words easier to understand. They are convenient and easy to use, especially in cases where it wouldn't be practical to use long terms over and over again. Researchers and professionals can share information more quickly and focus on the most important parts of their job by using abbreviations. Abbreviations also help people remember things, making it easier for them to understand complicated words and ideas. This can be very helpful when you are dealing with a lot of information, like when you are studying for a test or looking up information in scientific literature (Johnson, 2010).

But it's important to use nicknames carefully and with thought for the people you're writing for. Even though abbreviations can make communication easier, using them too much can make it hard for people who aren't familiar with the jargon or acronyms used in a certain area. So, it's important to find a good mix between using abbreviations and giving enough context or explanation for people who may not be familiar with the language. When using abbreviations in a scientific or professional setting, clarity and good communication should always come first.

They turn long words into short pictures, which makes it easier to share knowledge and saves valuable space in scientific writing. Abbreviations also help groups use the same language, which makes it easier for people from different countries to work together and share information. But it's important to use symbols wisely and think about the needs of the audience to make sure communication and understanding go smoothly. With careful use, abbreviations can continue to help advance scientific knowledge and make it easier for people in scientific and technical groups to talk to each other (Ehrenhauser, 2015).

10.3. COMMONLY USED ABBREVIATIONS IN CHEMISTRY

Because chemistry is a very complicated field of science, it uses symbols a lot to show and share information quickly. These abbreviations are now widely known and used by everyone in the field. They make it easier to talk about chemistry ideas, compounds, and processes. This talk will look at some of the most common chemistry abbreviations and explain what they mean, how they are used, and what they mean for scientific study and understanding.

pH is one of the most well-known scientific abbreviations. It stands for "potential of hydrogen." It is a way to measure how acidic or basic a material or solution is. The pH scale goes from 0 to 14. A value of 7 means that the substance is normal, a value below 7 means that it is acidic, and a value above 7 means that it is alkaline. pH is one of the most important ideas in chemistry. It helps us understand chemical reactions, biological processes, and systems in the world. It is used a lot in many different areas, such as biochemistry, environmental science, and research in the pharmaceutical industry (Johnson, 2010).

DNA is another well-known science acronym. It stands for "deoxyribonucleic acid." DNA is the inherited material found in all living things. It contains the genetic rules that control how they grow, work, and look. For many science fields, like genetics, molecular biology, and biotechnology, it is important to know how DNA is put together and what its properties are. James Watson and Francis Crick's finding of the double helix structure of DNA changed the way we think about genetics and led to a lot of progress in the field.

RNA, which stands for "ribonucleic acid," is also a very important part of many biological processes. It helps make proteins, controls genes and sends genetic information from one cell to another. DNA and RNA molecules have similar shapes, but their parts and roles are different. In recent years, there has been a lot of interest in the study of RNA because it could be used in gene therapy, drug creation, and to learn more about diseases.

ATP stands for "adenosine triphosphate," which is a molecule that gives live cells their main source of energy. It's a very important part of cellular processes like metabolism, muscle movement, and sending signals. ATP gives organisms the energy they need to do many different things, which is why it is so important to their health and life. You can't say enough about how important it is in biochemistry and cellular biology. Besides these specific examples, chemistry uses a lot of other names, each of which has its own meaning and purpose. For example, H_2O stands for water, CO_2 for carbon dioxide, and NaCl for sodium chloride, which is also known as table salt. These abbreviations make it easier to write down chemical molecules and let scientists talk to each other in a clear and concise way.

In chemistry, symbols are used for more than one reason. First, they make conversation easier by shortening long chemical names into easier-to-understand symbols. This makes science writing easier, so researchers can get their points across quickly and focus on the main ideas. Also, abbreviations make things clearer and easier to understand because they create a standard language that everyone in the field understands. They make it easier for scientists from different countries and backgrounds to talk to each other, removing possible language barriers and making sure that information flows smoothly. But it's important to use symbols in a way that makes sense for the people you're writing for. Even though abbreviations make it easier for scientists to talk to each other, they can be hard to understand for people who don't know the language or acronyms used in chemistry. Putting abbreviations in the right context and explaining them can help close this gap and make sure that more people can understand scientific results (Ehrenhauser, 2015).

In chemistry, abbreviations are very important because they make it easier to show and talk about complicated ideas, compounds, and processes. Abbreviations that are used a lot, like pH, DNA, RNA, and ATP, are now an important part of scientific study and understanding. They make it easier to talk to each other, make things clearer, and give everyone in the field a common language. But it's important to find a good mix between using abbreviations and giving enough background information for a wider audience to understand. By being used correctly, abbreviations continue to help advance scientific understanding and make it easier for people in the chemistry community to talk to each other.

10.4. LIMITATIONS OF THE IUPAC SYSTEM

The IUPAC is a group that is in charge of standardizing the names of chemical compounds and keeping them that way. The IUPAC system is a set of rules and guidelines that help scientists give new compounds they have found or made unique names. Even though the IUPAC method has been widely used and has helped a lot to organize and clarify the names of chemicals, it is not perfect and still has some problems (Johnson, 2010).

One of the biggest problems with the IUPAC method is that it is hard to understand. Naming chemicals can be complicated, and you need to know a lot about organic chemistry to do it right. The system is based on a hierarchical approach that takes into account things like the amount and arrangement of atoms and the stereochemistry of the compound. This can make things hard, especially for less experienced chemists or people from related areas who don't know much about how complicated the system is. Mastering the IUPAC rules could take a lot of time and work, which could lead to mistakes or flaws in naming.

The fact that the IUPAC system relies on formal names is another thing that holds it back. By taking a systematic method, the IUPAC rules try to give chemical compound names that are both unique and clear. But this can lead to names that are hard to say and remember because they are long and complicated. Systematic names often have prefixes, suffixes, and numbers that show how the combination is put together and what it is made of. Even though these names are accurate and clear from a chemistry point of view, they can be hard to remember and less useful in everyday life, especially in fields where compounds need to be found quickly and easily.

Also, the IUPAC method might have trouble naming newly discovered or made compounds that don't fit neatly into the rules for naming that are already in place. As science keeps making progress and new substances with different structures or qualities are found, the IUPAC system may need to change and make new rules to include them. This process of updating and adding to the system can take time, so there may be a delay between when a combination is found and when it gets a standard name. It can also lead to debates and conversations in the scientific community about how to name these new compounds in the best way. Also, the IUPAC system can be used and adopted in different ways in different places and scientific groups. Even though the IUPAC rules are known all over the world, some places or study groups may have their own rules or preferences for naming things that are different from the IUPAC system. This can lead to different names for the same compound, making it hard to come up with a title that everyone agrees on (Ehrenhauser, 2015).

Even though the IUPAC system is a good way to name chemical molecules and has made a big difference in the organization and clarity of chemical nomenclature, it has some problems and isn't perfect. There may be problems with the system due to its complexity, the possibility of long and hard-to-use systematic names, the need to change to include new combinations and differences in how names are used in different parts of the world. As the field of chemistry continues to change, it will be important for scientists to be able to communicate and understand each other if these problems are fixed and a balance is found between accuracy and usefulness in chemical nomenclature.

10.5. TRIVIAL NAMES

Trivial names in chemistry refer to the common names given to chemical compounds that are not based on any systematic naming conventions. These names are often derived from their physical or chemical properties, or from their historical uses. Trivial names are not standardized and can vary depending on the language or region, which can lead to confusion and errors in communication (Figure 10.3).

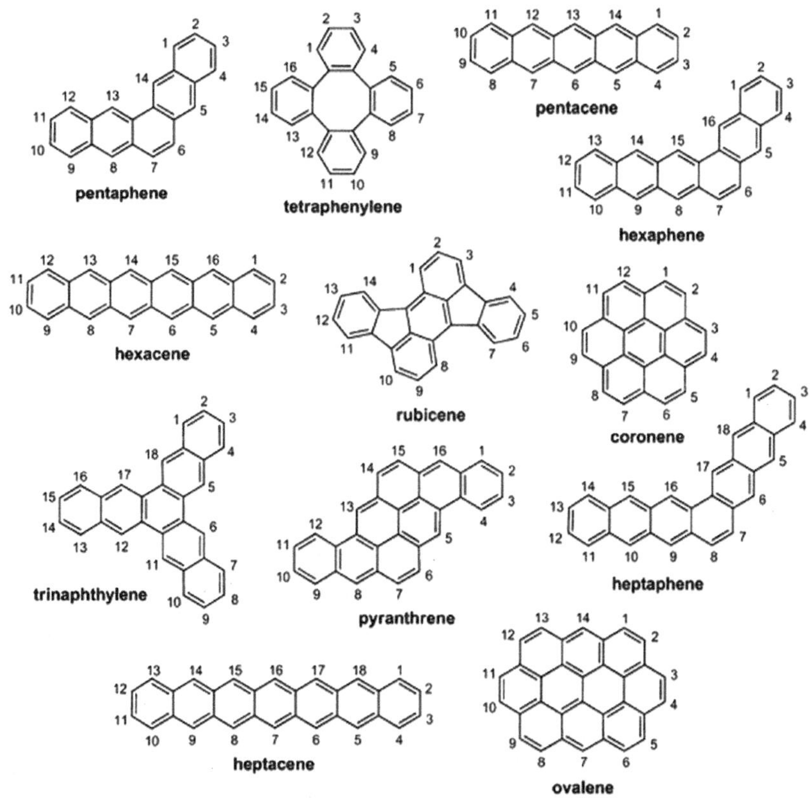

Figure 10.3. Trivial names of larger fused aromatic ring system.

Source: *Rasmussen, (2016).*

10.5.1. Trivial Names Based on History

In chemistry, trivial names are popular or non-systematic names for compounds that come from their history. These names often come from different places, like the name of the person who found the compound, its location, its physical qualities, or other things that describe it. Before systematic naming systems were made, trivial names were used to identify and talk about compounds. This was important in the history of chemistry. The name "aspirin," which stands for acetylsalicylic acid, is a well-known example of a trivial name. The name "aspirin" comes from the German word "*acetylspirsaure*," which is made up of the word "acetyl" for the acetyl group and "*Spirsaure*" for salicylic acid. The chemical was first made by Bayer in the late 1800s as a pain reliever and anti-inflammatory drug. The trivial name "aspirin" became so famous that it is still used a lot today, even though the scientific name follows the rules of the IUPAC (Ehrenhauser, 2015).

In the field of chemistry, there are several benefits to giving things trivial names. First of all, they can be used to refer to compounds in a way that is trivial and easy to remember. This makes it easier for scientists, students, and people, in general, to understand and talk about chemical substances. Names that aren't important often feel familiar, which makes them easier to talk about and remember. For example, "caffeine" is a common name for the natural stimulant 1, 3, 7-trimethylxanthine, which is found in coffee and tea. Also, trivial names can tell you important things about a compound's qualities, where it came from, or how it is meant to be used. For instance, the word "chloroform" refers to *trichloromethane*, which is a gaseous liquid with chlorine atoms in it. The name "chloroform" comes from the fact that it contains chlorine and comes from when it was used as an anesthetic. In the same way, "ethylene glycol" is just a trivial name for ethane-1,2-diol, which tells you about its chemical structure and how it works as a popular antifreeze.

But using trivial names in chemistry also has its downsides and problems. One big problem is that there isn't enough order and consistency. Names for trivial things are often arbitrary and can be different in different languages, places, and times. This can cause confusion, especially when different trivial names refer to the same substance or when two compounds have similar trivial names but have different structures or properties. Another problem with trivial names is that they could be misunderstood. The names of trivial compounds may not correctly describe their parts or structure, which can lead to misunderstandings or mistakes in scientific communication. Systematic names, like those set by the IUPAC, give a clear and exact way to talk about chemical compounds. This makes sure that scientific writing is consistent and clear (Johnson, 2010).

In order to fix these problems, the IUPAC method was made to give a standard and organized way to name chemical compounds. The IUPAC method uses rules and guidelines to give unique names to compounds based on their structure and how they work. The systematic names made by the IUPAC system make it easier for scientists from different fields and parts of the world to find, classify, and talk about chemistry compounds. In the end, trivial names in chemistry have played a big role in figuring out how to identify and talk about molecules based on their history. Even though they are trivial and easy to remember, trivial names have problems with standards, ambiguity, and accuracy. On the other hand, the IUPAC system gives a clear, consistent, and correct way to describe the chemical structure of a compound through its names. Chemical nomenclature has a long and interesting past, and it is still growing and changing. This lets scientists talk to each other about and study the wide range of chemical compounds.

10.5.2. Trivial Names Based on Geographical Factors

In chemistry, sometimes the names of trivial compounds are based on their location. This means that the name of a compound is linked to a certain area, location, or place. These names can tell us about the history of how the combination was found, where it came from, or how it is related to a certain place. Using geographical names in chemistry adds an interesting twist to the nomenclature and shows how scientific findings are linked to the places where they were made. One example of a trivial name based on location is "*Bergamotene*," which is a group of chemicals found in the essential oil of the bergamot orange (*Citrus bergamia*). Bergamotene molecules are sesquiterpenes that help give bergamot oranges their unique smell. The name "Bergamotene" is a clear reference to the citrus fruit's place of origin, which is the Italian region of Bergamo, where the fruit has been grown for a long time (Ehrenhauser, 2015).

In the same way, the word "Cubebene" comes from the plant *Cubeba officinalis*, which is also known as Java pepper or Cubeb. Cubebene is a natural chemical that is found in the essential oil of the Javanese plant Cubeb. The trivial name "Cubebene" refers to the area where the plant is most common. In some cases, the trivial name will include the name of the place or country where the compound was first found or isolated. For example, "Paris Green" is a name for copper(II) acetoarsenite, which is a green pigment that has been used in paints and poisons for a long time. The name "Paris Green" comes from the fact that it was used a lot in Paris in the 1800s.

Compounds can also get their names from where their minerals come from. For example, "turquoise" is a mineral and jewel that is well-known for its blue-green color. The name "turquoise" comes from the French word "turquois," which means "Turkish," because the stone was first brought to Europe through Turkish trade lines. People often think of turquoise as being from Persia (now Iran) or the American Southwest, where it has been found for hundreds of years. In addition to the names of individual compounds, geographical names can be used to group compounds that are similar to each other. For example, "Australian" or "Eucalyptol" refers to a group of chemicals called 1,8-cineole, which is often found in the essential oils of Australian eucalyptus species. The name "Australian" tells you where these chemicals come from and that they are linked to the smell of eucalyptus trees (Johnson, 2010).

Using geographical names in chemistry gives the compounds a cultural and historical meaning. It shows where the compounds come from in nature and where they were first found or are most common. These names not only help people understand where something is, but they also help scientists and business people identify, classify, and talk about chemicals. It's important to keep in mind, though, that while names based on geography can be helpful and serve as a form of cultural memory, they may not tell us much about the chemical structure or properties of the compounds. So, when scientific accuracy and clarity are very important, names defined by the IUPAC system are chosen because they are clear and follow a standard.

10.5.3. Trivial Names Based on Descriptive Factors

In chemistry, descriptive factors can also be used to come up with trivial names. For example, the name of a compound can be based on its physical qualities or functional groups. Most of the time, these descriptive names make it easy to find and explain compounds, especially when their systematic names are complicated or hard to understand. The use of detailed names makes nomenclature more useful and makes it easier for scientists and researchers to talk to each other. "Acetone," for example, is the name of the trivial and most well-known ketone molecule, which has the chemical formula CH_3COCH_3. The name "acetone" comes from the way the substance smells and tastes, which is similar to vinegar (acetic acid). The name "acetone" is a good way to describe the compound because it describes its qualities and makes it easier to find and identify (Ehrenhauser, 2015).

In the same way, "formaldehyde" is just a trivial name for the chemical substance with the formula CH_2O. The name "Formaldehyde" comes from the Latin word "formica," which means "ant." This is because the chemical has a strong smell that reminds people of ant waste. This name explains what the substance smells like, which makes it easy to find and recognize. Butane is another name that tells you something about what it is. It is a hydrocarbon with four carbon atoms and the chemical formula C_4H_{10}. Butane is an alkane, which means it has four carbon atoms, and the word "butane" is used to describe it. This name helps to tell it apart from other hydrocarbons with similar structures and traits but different names.

Some trivial names for mixtures are based on how they look or what they are made of. For example, "Red Phosphorus" is a popular name for the form of phosphorus that has a reddish-purple color, as opposed to "White Phosphorus," which has a yellowish-white color. The name "Red Phosphorus" draws attention to the visual difference between the two kinds of phosphorus and makes it easier to tell them apart. Trivial names that are descriptive can also be linked to certain functional groups or chemical qualities. For instance, "Hydrochloric Acid" is a popular name for the chemical compound HCl. The name "Hydrochloric Acid" tells us that it is acidic and that it contains a chloride ion (Cl^-). This name explains how the chemical compound works and helps to spot it in different situations (Johnson, 2010).

In addition to the names of individual compounds, detailed names can be used to put compounds into groups based on what they have in common. For example, "Alkyl Halides" is a general term for a group of compounds that have a halogen atom (like chlorine, bromine, or iodine) linked to an alkyl group. The name "Alkyl Halides" comes from the functional group that all of these compounds have in common. This makes it easy to recognize and classify these compounds. Names based on physical properties, functional groups, or other characteristics make it easy to find and explain compounds, especially the in lab and industrial settings where they are used every day. These names make it easier to talk about chemicals and to recognize and tell them apart based on their unique properties. But it's important to remember that a compound's name may not always tell you everything you need to know about its structure or how it behaves chemically. Because of this, the IUPAC system's systematic names are best for scientific writing and accurate chemical communication.

10.5.4. Benefits and Drawbacks of Trivial Names

One of the best things about trivial names is that they can make complicated chemical terms easier to understand for a bigger audience. Systematic names, which can be long and hard to say, are often easier to remember and say than trivial names. This makes communication easier, especially in non-scientific settings where exact chemical nomenclature may not be well understood by most people. Trivial names, like "aspirin" or "caffeine," have become part of daily language and are easy for most people to remember. This helps the public understand and be aware of certain chemical compounds. In science writing and discussion, trivial names can also be used as shorthand. Researchers and scientists often call chemicals by names that are easy to remember, especially when they are common or have been studied a lot. This makes it easier for scientists to talk to each other and lowers the need for long names that are used over and over again. It lets researchers share information quickly and clearly without having to use complicated chemical names. This is especially helpful in talks and presentations where brevity is key (Ehrenhauser, 2015).

Also, names that don't seem important can have historical or cultural meanings that tell us about the beginnings or properties of compounds. Some trivial names come from the name of the person who found the compound or from the place where it was found. This connects the name to the scientific past and memory. For example, "*Mendelian Genetics*" is called after the scientist Gregor Mendel, who was the first person to study how traits are passed down.

So, trivial names can be used to celebrate scientists and their work, which adds to the story and history of how science has progressed. But there are a few problems with giving insignificant names to things in science. One of the main worries is that they are not as clear and precise as systematic names. A compound's name may not tell you much about its chemical structure or its functional groups if it is just a trivial word. This lack of specificity can lead to confusion, especially when multiple compounds have the same trivial name or when a trivial name doesn't correctly describe the compound's properties. When names aren't clear, it can be hard to understand them and slow down scientific growth.

Another problem is that trivial names can be different in different places or languages, which can be confusing. Different countries may have different names for things that aren't important, which can cause confusion and miscommunication in foreign scientific collaborations. Also, using trivial names may make it harder to come up with a standard way to name things around the world, since different languages and cultures may have different trivial names for the same combination. This lack of unity can make it hard for scientists around the world to talk to each other and work together effectively. Also, names that seem like nothing can sometimes be wrong or false. They might spread false information or suggest things that science doesn't support. For example, "sweet oil" used to be a trivial name for a mixture of almond oil and olive oil, even though it wasn't sweet by nature. Such misleading names can make people believe things that aren't true, especially if the trivial name is used and accepted by many people without anyone questioning its scientific truth (Johnson, 2010).

To make up for these problems, the IUPAC has set up rules for naming chemical substances that make them clear and consistent. Based on a set of rules and guidelines, systematic names provide clear and consistent nomenclature that correctly describes the structure and composition of a compound. The IUPAC method makes sure that everyone understands the names of chemicals. This makes communication clear and prevents confusion.

10.5.5. Controversial Trivial Names

In chemistry, compounds with names that are offensive, misleading, or socially unacceptable are said to have "controversial trivial names." These names can cause problems in the scientific community and in society as a whole because they may reinforce stereotypes, offend certain groups, or make it harder for scientists to talk to each other. By looking at specific examples of controversial trivial names, we can learn more about the arguments about their use and see how important it is to use responsible and ethical naming practices (Ehrenhauser, 2015). One well-known example is the chemical called "adamantane." Even though the name itself might not be offensive, it has caused a lot of trouble because it comes from the word "adamantine," which means "unbreakable" or "unyielding." Some people say that using this name reinforces gender stereotypes because it makes people think that men are strong and resilient. This controversy shows how important it is to think carefully about the names given to compounds so that they don't reinforce stereotypes or make some groups feel left out.

Another example is "Redoxaflavin," which is an abbreviation for "reduction-oxidation-active flavin." People have said that the name "Redoxaflavin" could mislead and confuse people because it sounds like a flavin derivative that is involved in redox reactions. But "Redoxaflavin" does not have the structure or functions that are usually associated with flavins. This name is controversial because it could misrepresent the properties of the compound, which could lead to confusion and wrong conclusions.

Some silly names have also been called out for making people feel bad or being rude. For instance, "Negro Pepper" is a term for a spice made from the seeds of the *Xylopia aethiopica* tree. The word "Negro" in the name has gotten a lot of backlash and with good reason. It has racial overtones and uses old, offensive language. The controversy over these names shows how important it is for scientific nomenclature to be sensitive and inclusive, and to avoid terms that can make people or groups feel left out or offended (Johnson, 2010). In a similar way, the chemical compound "Indian Yellow" has been controversial because of its history and links to unethical practices. Indian Yellow was a color made from the urine of cows in India that were fed a certain diet. The extraction process was cruel to animals and raised moral questions. Even though the name may not be offensive on its own, the fact that it is linked to exploitative practices has led to discussions about whether or not it is okay to use such names and what the ethical implications of historical context are. These examples show how important it is to think carefully about even small names and what they might mean. To promote inclusion, respect, and accuracy in scientific communication, it is important to name compounds in a responsible and ethical way. Because of the problems with these names, scientists have talked about them and tried to change the rules for naming things so they don't have offensive or misleading associations. In response to these worries, the IUPAC has emphasized the need for responsible naming practices and encouraged scientists to think carefully about the cultural, historical, and social implications of meaningless names. The IUPAC system gives rules and guidelines to make sure that compound names are correct, helpful, and don't have any negative or offensive meanings (Ehrenhauser, 2015).

10.6. INTERNATIONAL STANDARDIZATION OF CHEMICAL NOMENCLATURE

Chemical nomenclature is an important part of how scientists talk to each other. It helps them identify, classify, and talk about chemical compounds in a standard way. Without a system for naming compounds that everyone agrees on, confusion and wrong interpretations can happen, slowing progress in many scientific fields. To deal with this problem, groups like the IUPAC have worked hard to set up and promote international chemical nomenclature standardization. This essay will talk about how important it is to use systematic naming conventions and look at what IUPAC is doing to help standardize things around the world. One of the main goals of standardizing chemical nomenclature is to make sure that scientists all over the world can talk to each other in a clear way. Researchers can avoid misunderstandings and work better together if they use the same rules and guidelines when naming compounds. Systematic nomenclature helps scientists find and tell the difference between different compounds, even if they work in different places or speak different languages. This is especially important in interdisciplinary research, where scientists from different fields work together to solve difficult problems (Johnson, 2010).

Adopting a system for naming things also makes scientific experiments more accurate and easy to repeat. Inconsistent or wrong names for compounds can cause confusion and mistakes in how data is interpreted, which slows the progress of science. With a standardized nomenclature system, researchers can easily find and re-create the compounds they use in their experiments. This makes sure that their results are accurate and that they can share them with the rest of the scientific community. This makes scientific research seem more reliable and trustworthy. Standardizing chemical nomenclature also makes it easier to find information and read scientific papers. Scientists often use databases and books to find out about specific compounds or research that is related to them. Without a standard way to name things, it can be hard to search for and find the right information. By naming things in a systematic way, researchers can easily look for compounds of interest, find relevant literature, and build on what they already know. This helps science move forward more effectively (Ehrenhauser, 2015).

The IUPAC is a very important part of global efforts to standardize. IUPAC is an international scientific organization that has been around since 1919. Its goal is to improve the chemical sciences and give advice on different parts of chemistry. One of its most important contributions is making and keeping up-to-date the IUPAC nomenclature rules, which give chemical compounds names that are consistent and logical. The IUPAC nomenclature rules cover many different ways to name chemicals, such as organic compounds, inorganic compounds, polymers, and biochemicals. These rules are updated often to take into account new discoveries, advances in science, and changes in language. They give rules about naming conventions, preferred suffixes and prefixes, stereochemistry, and other important things to keep compound names clear and accurate.

IUPAC not only sets rules for nomenclature but also works with other groups and national committees to promote the use of standard naming conventions around the world. This includes efforts to get scientific journals, educational materials, and chemical databases to use IUPAC's suggestions. By making it easier for scientists to work together and agree on things, IUPAC is a key player in getting systematic naming conventions used everywhere. International standardization in chemical nomenclature is important for more than just scientists. It also has real-world effects in places like businesses, government agencies, and patent offices. Standardized names make it easier for stakeholders to talk to each other and understand each other. This makes it easier to follow rules, protect intellectual property, and manage chemical substances. Standardized nomenclature also helps keep people safe by making sure chemicals are correctly identified and labeled. This makes mistakes and accidents less likely.

Standardizing chemical nomenclature has made a lot of progress, but there are still problems. It can be difficult to come up with naming rules for newly discovered compounds, complex organic molecules, and stereochemistry. Also, it can be hard to find a good balance between the need for names that are clear and tell you something and the need for names that are simple and easy to use. But ongoing research, collaboration, and involvement with the scientific community make it possible for naming conventions to keep getting better and more precise (Ehrenhauser, 2015).

10.7. CHALLENGES IN CHEMICAL NOMENCLATURE

Chemical nomenclature is important for scientists to be able to talk to each other in a clear and consistent way, but it also has some problems, especially when it comes to naming complex organic compounds, stereochemistry, and isomerism. These problems are caused by the complexity and size of organic chemistry. Compounds can have many different shapes and different ways of being arranged in space. In this section, we'll talk about these problems in more depth. Naming complex organic compounds is one of the most difficult parts of naming chemicals. Organic compounds can have more than one functional group, substituent, or cyclic structure. This makes the structures of these compounds long and complicated. It becomes hard to give these compounds names that are both clear and accurate. Systematic nomenclature rules, like those made by the IUPAC, tell us how to name complex compounds based on their structure, functional groups, and substituents. But the process can take a long time and requires a deep understanding of the principles of organic chemistry.

Stereochemistry is another big problem when it comes to naming chemicals. Stereoisomers are compounds that have the same molecular formula and are made up of the same atoms, but the way the atoms are arranged in space makes them different. Atoms can have different physical and chemical properties based on how they are arranged in space. This makes it very important to tell these isomers apart. To name stereoisomers, you have to say either how the substituents are arranged relative to each other or how they are arranged absolutely, using rules like the Cahn-Ingold-Prelog (CIP) rules. But it can be hard to show how the atoms are arranged in three dimensions in a two-dimensional chemical structure and to describe this in the name of the compound (Johnson, 2010).

Isomerism, which happens when two compounds have the same molecular formula but different structures, makes chemical nomenclature even more difficult. Isomers can be different in how they connect (structural isomers) or how they are set up in space (stereoisomers), or in both ways. To give an isomer a name, you have to tell it apart from other isomers and make sure that the name of the compound accurately reflects the differences in how it is made. In organic chemistry, for example, compounds can show positional isomerism, chain isomerism, or functional group isomerism, among other types of isomerism. Systematic naming rules are used to tell the difference between different isomers, but the process can be hard, especially when a compound can have more than one isomer (Ehrenhauser, 2015).

Bibliography

1. Achten, C., & Andersson, J. T., (2015). Overview of polycyclic aromatic compounds (PAC). *Polycyclic Aromatic Compounds*, *35*(2–4), 177–186.
2. Adam, C., Clark, S., & Ackland, G., & Crain, J., (1997). *Conformation-Dependent Dipoles of Liquid Crystal Molecules and Fragments from first Principles*. Scientific Figure on ResearchGate. Available from: https://www.researchgate.net/figure/Illustration-of-the-molecule-phenyl-benzoate-in-its-50-conformation-a-and-optimized_fig4_224746825 (accessed on 06 August 2023).
3. Adu-Gyamfi, K., Ampiah, J. G., & Appiah, J. Y., (2013). *Senior High School Chemistry Students' Performance in IUPAC Nomenclature of Organic Compounds, 8*(3), 68–75.
4. Arrowsmith, M., Hadlington, T. J., Hill, M. S., & Kociok-Köhn, G., (2012). Magnesium-catalyzed hydroboration of aldehydes and ketones. *Chemical Communications*, *48*(38), 4567–4569.
5. Bagal, D. B., & Bhanage, B. M., (2015). Recent advances in transition metal - catalyzed hydrogenation of nitriles. *Advanced Synthesis & Catalysis*, *357*(5), 883–900.
6. Bagger-Jørgensen, R., Meyer, A. S., Pinelo, M., Varming, C., & Jonsson, G., (2011). Recovery of volatile fruit juice aroma compounds by membrane technology: Sweeping gas versus vacuum membrane distillation. *Innovative Food Science & Emerging Technologies*, *12*(3), 388–397.

7. Beckett, M. A., Brellochs, B., Chizhevsky, I. T., Damhus, T., Hellwich, K. H., Kennedy, J. D., & Yerin, A., (2020). Nomenclature for boranes and related species (IUPAC recommendations 2019). *Pure and Applied Chemistry, 92*(2), 355–381.
8. Bharadwaj, N., (2016). *Measurement and Correlation of Aqueous Amine Solution Viscosities*. Scientific Figure on ResearchGate. Available from: https://www.researchgate.net/figure/6-Structure-of-aminesIntroduction-nomenclature-and-classification-of-aliphatic_fig5_337856622 (accessed on 06 August 2023).
9. Boyce, J. M., (2018). Alcohols as surface disinfectants in healthcare settings. *Infection Control & Hospital Epidemiology, 39*(3), 323–328.
10. Burkle-Vitzthum, V., Bounaceur, R., Paul-Marie, M., Montel, F., & Fusetti, L., (2011). *Mechanistic Model of the Thermal Evolution of n- and Iso-Alkanes (C1-C32) in Oils*. Scientific Figure on ResearchGate. Available from: https://www.researchgate.net/figure/Model-compounds-for-branched-alkanes-Fig-2-Examples-of-each-type-of-lumped-free_fig1_236845656 (accessed on 06 August 2023).
11. Burrows, A., Holman, J., Lancaster, S., Parsons, A., Overton, T., Pilling, G., & Price, G., (2021). *Chemistry3: Introducing Inorganic, Organic and Physical Chemistry*. Oxford University Press.
12. Canfield, P., Blake, I., Zheng-Li, C., & Luck, I., Krausz, E., Kobayashi, R., Reimers, J., & Crossley, M., (2018). A new fundamental type of conformational isomerism. *Nature Chemistry*. Scientific Figure on ResearchGate. Available from: https://www.researchgate.net/figure/Isomerism-hierarchy-IUPAC-classifications-of-isomerism-and-isomerization-types-with-an_fig1_325271357 (accessed on 06 August 2023).
13. Chen, H., Sun, S., Xi, H., Hu, K., Zhang, N., Qu, J., & Zhou, Y., (2019). Catalytic oxidative conversion of aldehydes into nitriles using $NH_3 \cdot H_2O/FeCl_2/NaI/Na_2S_2O_8$: A practical approach to febuxostat. *Tetrahedron Letters, 60*(21), 1434–1436.
14. Darwish, I. A., Darwish, H. W., Ali, A. M., & Almutairi, H. S., (2023). Spectrophotometric investigations of charge transfer complexes of tyrosine kinase inhibitors with iodine as a σ-electron acceptor: Application to development of universal high-throughput microwell assay for their determination in pharmaceutical formulations. *Medicina, 59*(4), 775. MDPI AG. Retrieved from: http://dx.doi.org/10.3390/medicina59040775.

15. Das, D. D., John, P. C. S., McEnally, C. S., Kim, S., & Pfefferle, L. D., (2018). Measuring and predicting sooting tendencies of oxygenates, alkanes, alkenes, cycloalkanes, and aromatics on a unified scale. *Combustion and Flame*, *190*, 349–364.

16. Dos, P. G. G., Wimmer, A., Spry, J., König, B., & Alabugin, I., (2019). CO_2 or SO_2: Should it stay, or should it go? *The Journal of Organic Chemistry*. Scientific Figure on ResearchGate. Available from: https://www.researchgate.net/figure/A-Historic-examples-of-utilizing-carboxylic-acids-and-sulfinates-as-synthetic-handles_fig1_332479827 (accessed on 06 August 2023).

17. Ehrenhauser, F. S., (2015). PAH and IUPAC nomenclature. *Polycyclic Aromatic Compounds*, *35*(2–4), 161–176.

18. Escott, et al., (2018). *Study of the Interaction of Anthocyanins with Phenolic Aldehydes in a Model Wine Solution.* Scientific Figure on ResearchGate. Available from: https://www.researchgate.net/figure/Phenolic-aldehydes-used-in-the-experimental-design-Names-and-aroma-profiles-are-listed_fig1_328970579 (accessed on 06 August 2023).

19. Estarellas, C., Frontera, A., Quiñonero, D., & Deyà, P., (2010). Can lone pair-π and cation-π interactions coexist? A theoretical study. *Central European Journal of Chemistry*. Scientific Figure on ResearchGate. Available from: https://www.researchgate.net/figure/Aromatic-compounds-A-C_fig1_227093551 (accessed on 06 August 2023).

20. Faisal, M., Saeed, A., Shahzad, D., et al., (2020). Recent developments and comparison of transformation strategies for organic halides to aldehydes and ketones. *Mol. Divers, 24*, 571–592. https://doi.org/10.1007/s11030-019-09956-1.

21. Fuse, S., Mifune, Y., & Takahashi, T., (2014). Efficient amide bond formation through a rapid and strong activation of carboxylic acids in a microflow reactor. *Angewandte Chemie International Edition*, *53*(3), 851–855.

22. Gheisari, A., (2017). *Novel Tools for Simultaneous Optogenetic Manipulation and Calcium Imaging in the Zebrafish Nervous System.* Scientific Figure on ResearchGate. Available from: https://www.researchgate.net/figure/Compounds-containing-fused-aromatic-rings-or-polycyclic-aromatic-hydrocarbons-PAH_fig3_318116264 (accessed on 06 August 2023).

23. Gohil, C., & Noolvi, M., (2021). Solvent-free synthesis of amide: A novel technique of green chemistry. *Asian Journal of Pharmaceutical and Clinical Research*. Scientific Figure on ResearchGate. Available from: https://www.researchgate.net/figure/Scheme-1-Conventional-method-of-amide-synthesis_fig1_351585166 (accessed on 06 August 2023).
24. Gomollón-Bel, F., (2019). Ten chemical innovations that will change our world: IUPAC identifies emerging technologies in Chemistry with potential to make our planet more sustainable. *Chemistry International*, *41*(2), 12–17.0
25. Goodarzi, B., Kalantar, Z., & Goodarzi, N., (2016). *Density Prediction of N-alkane and Cycloalkane Mixtures Using Calculated Molecular Descriptors and Levenberg–Marquardt Artificial Neural Network*, *4*(3), 57–68.
26. Gunawan, M. A., Hierso, J. C., Poinsot, D., Fokin, A. A., Fokina, N. A., Tkachenko, B. A., & Schreiner, P. R., (2014). Diamondoids: functionalization and subsequent applications of perfectly defined molecular cage hydrocarbons. *New Journal of Chemistry*, *38*(1), 28–41.
27. Gunes, A., Togan, M., Celik, F., & Cangul, I. N., (2019). Cut vertex and cut edge problem for topological graph indices. *Journal of Taibah University for Science*. Scientific Figure on ResearchGate. Available from: https://www.researchgate.net/figure/Symmetry-in-alkanes_fig1_337628843 (accessed on 06 August 2023).
28. Heller, S., McNaught, A., Stein, S., Tchekhovskoi, D., & Pletnev, I., (2013). InChI-the worldwide chemical structure identifier standard. *Journal of Cheminformatics*, *5*(1), 1–9.
29. Hintermann, L., (2010). Recent developments in metal-catalyzed additions of oxygen nucleophiles to alkenes and alkynes. *Cx Bond Formation*, 123–155.
30. Huang, X., & Ma, S., (2019). Allenation of terminal alkynes with aldehydes and ketones. *Accounts of Chemical Research*, *52*(5), 1301–1312.
31. Jarrell, J., (2008). *Active Metal Oxides and Polymer Hybrids as Biomaterials*. Scientific Figure on ResearchGate. Available from: https://www.researchgate.net/figure/Chemical-structure-of-first-second-and-third-degree-alcohols-useful-for-diluting_fig5_234478141 (accessed on 06 August 2023).

32. Johnson, A. D., (2010). An extended IUPAC nomenclature code for polymorphic nucleic acids. *Bioinformatics, 26*(10), 1386–1389.

33. Jones, R. G., Kitayama, T., Hellwich, K. H., Hess, M., Jenkins, A. D., Kahovec, J., & Wilks, E. S., (2016). Source-based nomenclature for single-strand homopolymers and copolymers (IUPAC recommendations 2016). *Pure and Applied Chemistry, 88*(10, 11), 1073–1100.

34. Kaur, R., Taheam, N., Sharma, A. K., & Kharb, R., (2013). Important advances on antiviral profile of chromone derivatives. *Research Journal of Pharmaceutical, Biological and Chemical Sciences, 4,* 79–96. Scientific Figure on ResearchGate. Available from: https://www.researchgate.net/figure/Chemical-structures-of-some-heterocyclic-compounds_fig2_260683286 (accessed on 06 August 2023).

35. Kim, J., Kim, H. J., & Chang, S., (2012). Synthesis of aromatic nitriles using nonmetallic cyano - group sources. *Angewandte Chemie International Edition, 51*(48), 11948–11959.

36. Kolar, M., Mozetic, M., Stana-Kleinschek, K., Fröhlich, M., Turk, B., & Vesel, A., (2015). *Covalent Binding of Heparin to Functionalized PET Materials for Improved Haemocompatibility.* Scientific Figure on ResearchGate. Available from: https://www.researchgate.net/figure/Schematic-of-4-chlorobenzaldehyde-binding-to-amino-groups_fig4_274321075 (accessed on 06 August 2023).

37. Korica, N., Freitas, M. P., Marin, G., & Thybaut, J., (2018). Mixture effect on alkane and cycloalkane hydroconversion over Pt/USY Catalyst. In: *Chemreactor-23 (Chemreactor-23).*

38. Krenske, E. H., Davison, E. C., Forbes, I. T., Warner, J. A., Smith, A. L., Holmes, A. B., & Houk, K. N., (2012). Reverse cope elimination of hydroxylamines and alkenes or alkynes: theoretical investigation of tether length and substituent effects. *Journal of the American Chemical Society, 134*(4), 2434–2441.

39. Lee, M., (2012). *Analytical Chemistry of Polycyclic Aromatic Compounds*. Elsevier.

40. Leigh, G. J., (2011). *Principles of Chemical Nomenclature: A Guide to IUPAC Recommendations*. Royal Society of Chemistry.

41. Li, Y., Lin, Y., Cheng, C., Hao, J., & Zhu, T., (2020). On the nature of nitrogen-containing groups in the SCR of NO over functionalized activated coke. *Waste and Biomass Valorization.* Scientific Figure

on ResearchGate. Available from: https://www.researchgate.net/figure/Main-types-of-oxygen-containing-and-nitrogen-containing-functional-groups-on-AC_fig5_327152243 (accessed on 06 August 2023).

42. Li, Z., Fang, M., LaGasse, M. K., Askim, J. R., & Suslick, K. S., (2017). Colorimetric recognition of aldehydes and ketones. *Angewandte Chemie*, *129*(33), 9992–9995.

43. Lippincott, D. J., (2018). *Allenes, Alkenes & Alkynes: My Piece of the π... in Water at Room Temperature*. Doctoral dissertation, University of California, Santa Barbara.

44. Liu, Y., Chase, H. M., & Geiger, F. M., (2019). Partially (resp. fully) reversible adsorption of monoterpenes (resp. alkanes and cycloalkanes) to fused silica. *The Journal of Chemical Physics*, *150*(7), 074701.

45. Lowe, D. M., (2012). *Extraction of Chemical Structures and Reactions from the Literature*. Doctoral dissertation, University of Cambridge.

46. Malapit, C. A., Borrell, M., Milbauer, M. W., Brigham, C. E., & Sanford, M. S., (2020). Nickel-catalyzed decarbonylative amination of carboxylic acid esters. *Journal of the American Chemical Society*, *142*(13), 5918–5923.

47. Man, A., Gaz, S. A., & Mare, A., & Berța, L., (2017). Effects of low-molecular-weight alcohols on bacterial viability. *Revista Romana de Medicina de Laborator*, *25*, 335–343. Scientific Figure on ResearchGate. Available from: https://www.researchgate.net/figure/Structure-of-the-used-alcohols_tbl1_320624607 (accessed on 06 August 2023).

48. Mcnulty, J., Krishnamoorthy, V., & Robertson, A., (2008). *ChemInform Abstract: Direct Formation of Esters and Amides from Carboxylic Acids Using Diethyl Chlorophosphate in Pyridine*. Scientific Figure on ResearchGate. Available from: https://www.researchgate.net/figure/Synthesis-of-various-esters-and-amides_tbl1_244235787 (accessed on 06 August 2023).

49. Montgomery, M. J., Zhu, J., Pfefferle, L. D., & McEnally, C. S., (2021). Amines have lower sooting tendencies than analogous alkanes, alcohols, and ethers. *Combustion and Flame*, *227*, 335–345.

50. Mu, Y., Nguyen, T., Koh, M. J., & Schrock, R., & Hoveyda, A., (2019). E - and Z -, di- and tri-substituted alkenyl nitriles through catalytic cross-metathesis. *Nature Chemistry*. Scientific Figure on ResearchGate. Available from: https://www.researchgate.net/figure/

Biologically-active-compounds-with-an-alkenyl-nitrile-or-a-related-moiety-a_fig1_332133154 (accessed on 06 August 2023).

51. NguyenHuynh, D., (2018). Accurate modeling of multiphase behavior of aqueous systems. I. Alkanes, alkenes, cycloalkanes, alcohols, aromatics. *Fluid Phase Equilibria*, *473*, 201–219.

52. Nugent, et al., (2021). A 2000 to 2020 practitioner's guide to chiral amine based enantioselective aldol reactions: Common ketone substrates, best methods, in water reaction environments, and defining nuances. *European Journal of Organic Chemistry*. Scientific Figure on ResearchGate. Available from: https://www.researchgate.net/figure/Ketones-summarized-in-this-review-the-corresponding-tables-provide-the-best-reaction_fig1_354304816 (accessed on 06 August 2023).

53. Ouédraogo, I., Boulvin, M., & Flammang, R., Gerbaux, P., & Bonzi-Coulibaly, Y., (2009). *Conversion of Natural Aldehydes from Eucalyptus Citriodora, Cymbopogon Citratus, and Lippia Multiflora into Oximes: GCMS and FT-IR Analysis*. Scientific Figure on ResearchGate. Available from: https://www.researchgate.net/figure/Structures-of-various-aldehydes-studied-and-names-of-oxime-and-nitrile-homologues_tbl1_26840743 (accessed on 06 August 2023).

54. Patron, E., (2021). *Exploring the Role That Visual Representations Play When Teaching and Learning Chemical Bonding: An Approach Built on Social Semiotics and Phenomenography*. Scientific Figure on ResearchGate. Available from: https://www.researchgate.net/figure/Different-visual-representations-of-a-methane-molecule-a-a-space-filling-molecular_fig4_357151895 (accessed on 06 August 2023).

55. Pérez, D., Peña, D., & Guitián, E., (2013). Aryne cycloaddition reactions in the synthesis of large polycyclic aromatic compounds. *European Journal of Organic Chemistry*, *2013*(27), 5981–6013.

56. Qiu, D., & Sulston, K., (2016). *Electron Transmission through Modified Benzene*. Scientific Figure on ResearchGate. Available from: https://www.researchgate.net/figure/Modified-benzene-molecule-showing-para-meta-and-ortho-connections-to-leads_fig5_305007193 (accessed on 06 August 2023).

57. Rasmussen, S. C., (2016). The nomenclature of fused-ring arenes and heterocycles: A guide to an increasingly important dialect of organic chemistry. *ChemTexts 2, 16*(2016). https://doi.org/10.1007/s40828-016-0035-3.

58. Riss, P. J., Lu, S., Telu, S., Aigbirhio, F. I., & Pike, V. W., (2012). CuI - catalyzed 11C carboxylation of boronic acid esters: A rapid and convenient entry to 11C - labeled carboxylic acids, esters, and amides. *Angewandte Chemie International Edition, 51*(11), 2698–2702.
59. Sahoo, R., Das, A., Gaur, M., Dey, S., Sahoo, S., & Subudhi, E., (2019). Bacteria for butanol production: Bottlenecks, achievements, and prospects. *Journal of Pure and Applied Microbiology*. Scientific Figure on ResearchGate. Available from: https://www.researchgate.net/figure/Isomers-of-Butanol-and-their-chemical-structures_fig1_336152895 (accessed on 06 August 2023).
60. Saladino, R., Brucato, J., & De Sio, A., Botta, G., Pace, E., & Gambicorti, L., (2011). Photochemical synthesis of citric acid cycle intermediates based on titanium dioxide. *Astrobiology*. Scientific Figure on ResearchGate. Available from: https://www.researchgate.net/figure/Long-chain-and-other-carboxylic-acids-13-19-and-heterocycles-20-21-obtained-by_fig4_51728365 (accessed on 06 August 2023).
61. Schmiermund, T., (2023). Hydrocarbons. In: *The Chemistry Knowledge for Firefighters*. Springer, Berlin, Heidelberg. https://doi.org/10.1007/978-3-662-64423-2_44.
62. Scotti, C., & Barlow, J. W., (2022). Natural products containing the nitrile functional group and their biological activities. *Natural Product Communications, 17*(5). https://doi.org/10.1177/1934578X221099973.
63. Şendur, G., (2012). Prospective science teachers' misconceptions in organic chemistry: The case of alkenes. *Journal of Turkish Science Education, 9*(3), 186–190.
64. Shahidi, F., Varatharajan, V., Oh, W., & Peng, H., (2019). Phenolic compounds in agri-food by-products, their bioavailability and health effects. *Journal of Food Bioactives*. Scientific Figure on ResearchGate. Available from: https://www.researchgate.net/figure/Structure-of-chlorogenic-acid_fig6_332509669 (accessed on 06 August 2023).
65. Smith, B., (2012). *Bridged Aromatic Compounds*. Elsevier.
66. Sribalan, R., Sangili, A., Banuppriya, G., & Padmini, V., (2017). An efficient synthesis of nitrile, tetrazole and urea from carbonyl compounds. *New Journal of Chemistry, 41*(9), 3414–3421.
67. Sukhorukov, A. Y., (2020). Catalytic reductive amination of aldehydes and ketones with nitro compounds: new light on an old reaction. *Frontiers in Chemistry, 8*, 215.

68. Szatylowicz, H., Wieczorkiewicz, P. A., & Krygowski, T. M., (2022). Aromaticity concepts derived from experiments. *Sci., 4*(2), 24. MDPI AG. Retrieved from http://dx.doi.org/10.3390/sci4020024.

69. Taylor, R., Cole, J., Cosgrove, D., Gardiner, E., Gillet, V., & Korb, O., (2012). Development and validation of an improved algorithm for overlaying flexible molecules. *Journal of Computer-Aided Molecular Design*. Scientific Figure on ResearchGate. Available from: https://www.researchgate.net/figure/Example-overlay-of-nitrile-carbonyl-and-phosphate-acceptors-Each-acceptor-is-shown-with_fig1_224854999 (accessed on 06 August 2023).

70. Tevis, D. S., Flores, S. R., Kenwood, B. M., Bhandari, D., Jacob, 3rd. P., Liu, J., & De Jesús, V. R., (2021). Harmonization of acronyms for volatile organic compound metabolites using a standardized naming system. *International Journal of Hygiene and Environmental Health, 235*, 113749.

71. Vančik, H., (2022). Alkanes, composition, constitution, and configuration. In: *Basic Organic Chemistry for the Life Sciences*. Springer, Cham. https://doi.org/10.1007/978-3-030-92438-6_1.

72. Vančik, H., (2022). Alkenes and alkynes. In: *Basic Organic Chemistry for the Life Sciences* (pp. 43–61). Cham: Springer International Publishing.

73. Vančik, H., (2022). Alkenes and alkynes. In: *Basic Organic Chemistry for the Life Sciences*. Springer, Cham. https://doi.org/10.1007/978-3-030-92438-6_4.

74. Vohlídal, J., Wilks, E. S., Yerin, A., Fradet, A., Hellwich, K. H., Hodge, P., & Stepto, R. F., (2012). Terminology and nomenclature for macromolecular rotaxanes and pseudorotaxanes (IUPAC Recommendations 2012). *Pure and Applied Chemistry, 84*(10), 2135–2165.

75. Walker, M. J., Burns, D. T., Axford, I., & Moss, G. P., (2021). Cannabinoids—A tutorial review psychoactivity, regulation, common and IUPAC nomenclature, structures, and abbreviations in relation to cannabidiol (CBD) products. *J. Assoc. Publ. Analysts, 49*, 1–28.

76. Wang, L., Zhang, T., Sun, W., He, Z., Xia, C., Lan, Y., & Liu, C., (2017). C–O functionalization of α-oxyboronates: A deoxygenative gem-diborylation and gem-silylborylation of aldehydes and ketones. *Journal of the American Chemical Society, 139*(14), 5257–5264.

77. Wang, M. X., (2015). Enantioselective biotransformations of nitriles in organic synthesis. *Accounts of Chemical Research, 48*(3), 602–611.
78. Wang, Z., Wang, X., & Nishihara, Y., (2020). Nickel or palladium - catalyzed decarbonylative transformations of carboxylic acid derivatives. *Chemistry–An Asian Journal, 15*(8), 1234–1247.
79. Werkmeister, S., Junge, K., & Beller, M., (2014). Catalytic hydrogenation of carboxylic acid esters, amides, and nitriles with homogeneous catalysts. *Organic Process Research & Development, 18*(2), 289–302.
80. Wu, M., Wang, B., Hu, Q., Wang, L., & Zhou, A., (2018). The synthesis process and thermal stability of V2C MXene. *Materials.* Scientific Figure on ResearchGate. Available from: https://www.researchgate.net/figure/Abbreviations-and-the-corresponding-full-names-of-chemical-substances_tbl1_328587259 (accessed on 06 August 2023).
81. Yan, X., & Harutyunyan, S., (2019). Catalytic enantioselective addition of organometallics to unprotected carboxylic acids. *Nature Communications.* Scientific Figure on ResearchGate. Available from: https://www.researchgate.net/figure/State-of-the-art-in-conjugate-additions-to-unsaturated-carboxylic-acids-and-potential_fig1_334768872 (accessed on 06 August 2023).
82. You, L., Pescitelli, G., Anslyn, E. V., & Di Bari, L., (2012). An exciton-coupled circular dichroism protocol for the determination of identity, chirality, and enantiomeric excess of chiral secondary alcohols. *Journal of the American Chemical Society, 134*(16), 7117–7125.
83. Yuanxun, W., Du, Y., & Huang, N., (2018). A survey of the role of nitrile groups in protein-ligand interactions. *Future Medicinal Chemistry.* Scientific Figure on ResearchGate. Available from: https://www.researchgate.net/figure/Nitrile-groups-forming-hydrophobic-interactions-A-At-left-improved-potency-upon_fig4_329452578 (accessed on 06 August 2023).
84. Zhen-Hua, Z., & Xiao-Yang, D., & Xuan-Yi, D., & Qiang-Shuai, G., & Zhong-Liang, L., & Xin-Yuan, L., (2019). Copper-catalyzed enantioselective Sonogashira-type oxidative cross-coupling of unactivated C(sp^3) –H bonds with alkynes. *Nature Communications, 10,* 5689. 10.1038/s41467-019-13705-1. Scientific Figure on ResearchGate. Available fromhttps://www.researchgate.net/figure/Substrate-scope-of-alkynes-Standard-conditions-1aa-02mmol-alkyne-04mmol-CuI_fig3_337920181 (accessed on 06 August 2023).

INDEX

B

C

D

E

N

O

P

T

U

V

W

X